T0358880

Routledge Revivals

The Meaning of the Concept of Probability in Application to Finite Sequences

First published in 1990, this is a reissue of Professor Hilary Putnam's dissertation thesis, written in 1951, which concerns itself with *The Meaning of the Concept of Probability in Application to Finite Sequences* and the problems of the deductive justification for induction. Written under the direction of Putnam's mentor, Hans Reichenbach, the book considers Reichenbach's idealization of very long finite sequences as infinite sequences and the bearing this has upon Reichenbach's pragmatic vindication of induction.

The Meaning of the Concept of Probability in Application to Finite Sequences

Hilary Putnam

Routledge
Taylor & Francis Group

First published in 1990
by Garland Publishing

This edition first published in 2011 by Routledge
2 Park Square, Milton Park, Abingdon, Oxon, OX14 4RN

Simultaneously published in the USA and Canada
by Routledge
711 Third Avenue, New York, NY 10017

Routledge is an imprint of the Taylor & Francis Group, an informa business

Publisher's Note
The publisher has gone to great lengths to ensure the quality of this reprint but
points out that some imperfections in the original copies may be apparent.

Disclaimer
The publisher has made every effort to trace copyright holders and welcomes
correspondence from those they have been unable to contact.

A Library of Congress record exists under ISBN: 0824032098

ISBN 13: 978-0-415-68794-2 (hbk)
ISBN 13: 978-0-203-35744-6 (ebk)

Harvard Dissertations in Philosophy ——

Edited by

**Robert Nozick
Arthur Kingsley
Porter Professor of
Philosophy
Harvard University**

A GARLAND SERIES

The Meaning of the Concept of Probability in Application to Finite Sequences

Hilary Putnam

GARLAND PUBLISHING
NEW YORK & LONDON
1990

Library of Congress Cataloging-in-Publication Data

Putnam, Hilary

The meaning of the concept of probability in application to
finite sequences/ Hilary Putnam.

p.   cm. — (Harvard dissertations in philosophy)

Thesis (Ph.D.)—University of California, Los Angeles, 1951.

Includes bibliographical references.

ISBN 0-8240-3209-8

1. Probabilities. 2. Sequences (Mathematics) 3. Induction
(Logic) I. Title. II. Series.

QA273.P87   1990

519.2—dc20   89-49419

All volumes printed on acid-free, 250-year-life paper
Manufactured in the United States of America

Design by Julie Threlkeld

INTRODUCTION SOME YEARS LATER

Looking at my dissertation 38 years later reacquainted me with a young philosopher who was strangely somehow myself. Although my views have inevitably changed since I wrote it, I am glad to see that, even as a graduate student writing a thesis in philosophy of science, I took the trouble to give my technical problem a historical setting, and to explain its wider philosophical significance. I did not believe (and I do not believe now) that the use of analytical methods in philosophy is incompatible with an interest in the big questions. The big questions with which the dissertation deals are the justification of induction and the meaning of probability, and they were central in the life work of my teacher and thesis advisor, or as the Germans so well say, my *Doktorvater* , Hans Reichenbach.

Reichenbach was neither the first nor the last philosopher to believe that one can meet Hume's sceptical challenge, and provide a deductive justification for induction, but his was by far the most important attempt (even if I no longer believe that he succeeded) and the attempt from which one can learn the most. Reichenbach conceded, as one must concede, that Hume was partly right: one cannot hope to give a deductive proof that induction will succeed in the long run.[1] But what Hume's argument failed to rule out, according to Reichenbach, was the possibility of a deductive proof that induction will succeed *if* any method at all can succeed. And that is what Reichenbach thought he had provided.

In 1951 there was a great deal of unclarity about the nature of probability, and there continues to be a great deal of unclarity today. Many mathematicians did not understand that providing axioms for probability theory is not the same thing as interpreting the notion, while others failed to distinguish between abstract measure theory and interpreted probability theory. Reichenbach, following the lead of C.S. Peirce and von Mises, identified probability with the relative frequency of an attribute in a finite population (which Reichenbach thinks of as a finite sequence), or the limit of the relative frequency of the attribute in an infinite sequence. This put him in conflict with his good friend Rudolf Carnap who at that time followed Keynes in thinking of probability as a primitive logical notion.[2] For Reichenbach this talk of a primitive logical notion was little better than sheer mysticism.

Although Reichenbach held that probability could be either frequency in a finite population or the limit of frequency in an infinite one, in his great work on the theory of probability[3] he idealized finite populations as infinite ones, because his proofs of some of the fundamental theorems (e.g. Bernoulli's theorem) required that the population be infinite. One of the tasks of my dissertation was to show how one could develop the theory of probability using only finite populations, and this task gives the thesis its title. A technical device which proved to be of great usefulness in this connection was to define "the i th member of the sequence" in such a way that for $i > N$ (the length of the sequence), the i th member of the sequence is just i modulo N.[4] This has the consequence that "the i th member of the sequence" is defined for every value of the index i. In the dissertation,

2

I give axioms for such a finite form of the theory of probability and derive the major theorems. (This is certainly the most boring part of the dissertation.)

It was, however, in the course of this relatively boring work that I discovered the theorem that gave the dissertation its real interest in my own eyes as well as in the eyes of its readers. To our surprise, it turned out that the axioms I had given as the natural modifications of Reichenbach's axioms for the finitary context that I was dealing with, admitted of only one interpretation! namely the relative frequency interpretation.

Looking back on that result after so many years of experience with other ways of thinking than Reichenbach's, the result is less amazing, but it remains somewhat surprising. What Reichenbach and I both took for granted in our axioms, and what Keynes and Carnap would not have accepted, is that probability is an *extensional* concept. That means that if two attributes F and G are coextensive (apply to exactly the same things), then the probability that a member of the populations is an F must equal the probability that it is a G. Evident as this sounds, it is rejected (and must be rejected) by those who favor other interpretations of probability. Other interpretations hold that probability is an *intensional* notion, that is, that if F and G are *logically equivalent* attributes then the probability that a member of the population is an F equals the probability that it is a G, but not that it is extensional. The fact that my axioms include an axiom of extensionality is what makes my result not totally unbelievable; but Reichenbach's axioms also include an axiom of extensionality, and no similar result

3

seems to be provable for the infinite case. Even if one is not wedded to the frequency interpretation as the *only* interpretation of probability, the fact that in the finitary case it is the unique extensional interpretation remains, perhaps, of philosophical interest.

The justification of induction and the finite case

What led me to concentrate on the finite case was not an interest in probability theory as such. Rather I was interested, as I still am, in Reichenbach's justification of induction, and one major difficulty (pointed out by Reichenbach himself) with that justification involved the issue of the finite versus the infinite. Reichenbach' s justification involved an idealized immortal inquirer who continues trying to determine the limit of the relative frequency in one or more infinite sequences. (According to Reichenbach) relying on induction means following the policy of positing that the limit of the relative frequency of an attribute is approximately equal to the relative frequency among the cases so far observed.

Reichenbach showed that by employing induction one will eventually make correct posits (and no incorrect ones, once the "point of convergence" is reached) *provided* the relative frequency approaches a definite limit at all. (The other part of Reichenbach's justificatory argument is that successful prediction is impossible if limits to the relative frequency do not exist.) While a similar argument had been given earlier by Peirce, Peirce's arguments seem to assume that the sequences we encounter in nature have certain randomness properties, and thus smuggles in an empirical assumption.

The difficulty is that we only know that the immortal predictor will *sooner or later* make true predictions (if there is a limit); but there is no finite N such that we can say in advance that he will make true predictions before he reaches the Nth member of the sequence. If the "infinite sequences" that Reichenbach speaks of are really only very long finite sequences, if treating them as infinite is an "idealization", then does Reichenbach's whole justification not collapse? What relevance does Reichenbach's argument have for us all-too-mortal predictors?

It was this objection that I tried to defuse in the final chapter of the thesis, and it was in order to set up this problem in proper Reichenbachian fashion that I formalized the whole theory of probability as a theory dealing with finite sequences.

Afterthoughts on induction

As I now see it, there is a much more serious flaw in Reichenbach's argument than "the problem of the finite case". The problem, which was first pointed out by Nelson Goodman, is the problem of the *consistency* of induction. Goodman's famous paradox involving the attribute "grue"[5] is only a special case of the fact that if we follow Reichenbach's Rule of Induction to the letter, we will make logically contradictory forecasts. For example, suppose I have observed that a certain process produces groups of dots, and that the numbers in the successive groups observed so far have been 1,2,3,5 and 8. Should I predict that I will observe a group with 13 dots next (on the ground that 1,2,3,5,and 8 are Fibonacci numbers, and 13 is the next Fibonacci number)? Or should I predict that only groups of 1,2,3,5 and

5

8 dots will be observed in the future? Either prediction can be described as an "induction"; in the one case, I am positing that the hypothesis "The number of dots in the n th group is always the n th Fibonacci number" will continue to be confirmed, and in the other case I am positing that the hypothesis "The number of dots in a group is always either 1,2,3,5 or 8" will be. If I follow the Rule of Induction with respect to the attributes "is in accordance with the first hypothesis" and "is in accordance with the second hypothesis" I will make logically contradictory predictions.

Reichenbach mentions this problem in the English edition of his *Theory of Probability*, and dismisses it with the puzzling statement that "with respect to consistency, inductive logic differs intrinsically from deductive logic; it is consistent not *de facto* but *de faciendo*, that is, not in its actual status, but in a form to be made." (p. 450) I take it that what Reichenbach means is that one must make an arbitrary choice[6] of attributes to get started in induction: if I choose to project Goodman's funny predicate "grue" instead of the more normal "green", experience will show me the error of my ways in the fullness of time. If I project the Fibonacci numbers, then I must not also project the other hypothesis; but if the 6th group of dots turns out to have 3 members, then I will have to abandon the hypothesis that all the Fibonacci numbers appear in order, and I may then project the other (still unfalsified) hypothesis. While this sounds plausible, it does not really suffice.

The reason it does not suffice is that there is no guarantee that the correct hypothesis will *ever* be tried. If I choose to project a

6

hypothesis which is still unfalsified but incompatible with the correct hypothesis, then in time experience will show me that that hypothesis is not true (at least in the case of simple universal hypotheses like the ones mentioned). But then I will have to decide what hypothesis to project next; and there is no guarantee that the as-yet-unfalsified hypothesis I choose to project at *this* stage will not also be incompatible with the correct hypothesis (which means that I will once again be "blocked" from also projecting the correct hypothesis, on pain of making inconsistent predictions) There are, after all, infinitely many hypotheses which have not been falsified by any given finite amount of evidence, and which are incompatible with one another and with the correct hypothesis. And making higher-level inductions will not eliminate the wrong ones unless I choose to project the correct hypothesis at the meta-level; for the logical situation is the same at all levels.

Reichenbach might reply that unless at some stage the correct hypothesis occurs to us as one we should test, then no method will enable us to make correct predictions. (This is, at any rate, what Peirce says.) Again, this sounds plausible, but does not suffice.

It doesn't suffice because it is not necessarily impossible to make successful predictions without the correct hypothesis spontaneously "occurring" to one. For example, we might build a machine which suggests hypotheses at random (or which constructs only hypotheses which would never occur to a sane human scientist) and test those instead of the hypotheses which "occur to" a human scientist. And there is a logically possible world in which the correct hypothesis would be found by such a method, and would never be found by successively testing the hypotheses that naturally "occur to" us. That successful prediction is impossible unless the correct hypothesis is one which will sooner or later occur to a human scientist is not a logical truth, but an empirical principle---one which cannot itself be known to be true without making an induction.

Reichenbach considers an objection like this in *The Theory of Probability* . His answer is that if some people were to show us that they can make successful predictions by employing a bizarre method (Reichenbach imagines a consistently successful fortune teller), then those who use induction will verify that relying on the bizarre method is a good idea using *their* method---induction! This idea behind this answer can be elaborated to show that *if any other method is actually used and succeeds, induction will eventually succeed* , but this is less than Reichenbach needs to show. What Reichenbach claimed is that if any other method *can* succeed, then induction will succeed, and that is very different. The difficulty is that what methods are actually tried depends on what people believe. *We* believe that induction will work-- where by induction we mean, let us say, applying the Rule of Induction to those hypotheses that actually occur to scientists (presumably, the hypothesis that the bizarre method works would occur to scientists, if people actually used it and got good results long enough). In this connection, one can also have subordinate rules saying that some hypotheses are to be tried before others---ones which are more falsifiable, cost less to test, etc.[7] The prevalence of belief in induction (however "induction" be restricted, to avoid the making of inconsistent predictions) certainly effects what methods and hypotheses occur to people who are scientific and even to people who are non-scientific or anti-scientific---effects it in a variety of ways. If we did not believe in induction, then it might be that methods would get tried---*and would succeed* --- that will never get tried (and hence never get confirmed) in the actual world. In short, it could be logically (and even practically) possible to make successful predictions, even though we (using induction) will never make them and no one else will ever make them either (although people *would have made them* if it had not been for the faith in induction). No only is there no deductive proof that induction will succeed, as Reichenbach concedes; there is also no deductive guarantee that induction will succeed *provided* successful prediction is possible at all. Every sound argument for relying on induction relies---and *must* rely--- at some point on beliefs about the general course of the world.

Induction and recursion theory

The flaw in Reichenbach's attempted justification of induction--- the difficulty that the Rule of Induction is inconsistent unless restricted---can be viewed in another way. If we could arrange *all* testable hypotheses in a single list, and then proceed to test hypotheses in the order in which they occur in our list, perhaps we could ensure that *if* there is a correct testable hypothesis, then sooner or later it will get tested (at least if there is an "immortal inquirer" or an immortal community of inquirers; the problem of the finite case arises once more here.)

I can illustrate what this means by returning to my example of the process which generates groups of dots. Let us suppose that each hypothesis that we are interested in says that, for all n, the number of dots in the n th group observed will be $f(n)$, where f is a mathematical function, defined for at least some positive integral values of n. According to Church's Thesis, such a function is calculable by human beings if and only if it is partial recursive. If we suppose that a hypothesis of this form is "testable" just in case the function f is calculable (which is already problematic, by the way[8]), then constructing an infinite list of all the testable hypotheses (in this case) involves constructing an infinite list of all the partial recursive functions, and this is something mathematicians know how to do. This does not quite solve the "consistency" problem, but it can, in fact, be shown that there is an effective rule for making inductions (one that a computing machine could be programmed to follow) which will ensure that if a hypothesis in the infinite list is correct, it will eventually be projected (and not subsequently abandoned). But again an empirical hypothesis is involved in saying that *all* testable hypotheses are in our list: Church's Thesis is itself an empirical hypothesis about the calculating powers of human beings, one which it needs induction to verify. If we use a method which takes account only of recursive hypotheses, then if we live in a world in which human beings have the ability to calculate nonrecursive functions, and the true hypothesis involves a function of this kind, then we will never confirm the true hypothesis using a scheme of the kind just described. In sum: Reichenbach's claim, that using induction one must confirm the true hypothesis eventually if there is a true testable hypothesis, is right only if induction is applied to *all* the hypotheses we could possibly test; but then we will make inconsistent predictions, so the fact that some of our predictions will be "true" means nothing. And if we remove the inconsistency by restricting or ordering the hypotheses to be tested in some way, then Reichenbach's argument loses its deductive validity.

Even if Reichenbach's (and my younger self's) aim of somehow deductively justifying induction turns out to be an unattainable one, the discussion of Reichenbach's argument leads into profound depths. We have learned and will continue to learn by exploring the depths. What we will not find (I predict) is the kind of deductive refutation of the sceptic that so many philosophers dreamed of. In a sense, my present stance is Wittgensteinian; like Wittgenstein, I believe that our lives and our knowledge do not rest on apodictic reason[9]. But they are our lives and it *is* knowledge---knowledge humanly speaking. Where I perhaps differ with Wittgenstein is in finding the attempts of the foundationalists of permanent value nonetheless.

HILARY PUTNAM
SEPTEMBER 3, 1989

ENDNOTES

1 However there is an error or exaggeration in Hume's scepticism which was pointed out by Kant: while Hume was right, that successful prediction of the particular sequence of observable qualities in time may elude us, the very notion of a world in which it makes sense to say this---a world in which we can speak of *time* , or of observable qualities---presupposes a good deal of regularity. Hume characteristically assumes that qualities are independent of *laws* ---but this is a mistake.

2 Cf. *Logical Foundations of Probability* (University of Chicago, 1950), especially chapter II. Carnap conceded that probability *sometimes* means frequency. His doctrine was that there are two notions of probability; but he gave pride of place to the so-called logical notion.

3 *Wahrscheinlichkeitslehre* (Leiden, 1935). English trans. with new additions: *The Theory of Probability* (University of California, 1940).

4 I.e., the remainder one gets when one divides *I* by N.

5 "Now let me introduce another predicate less familiar than 'green'. It is the predicate 'grue' and it applies to all things examined before *t* just in case they are green but to other things just in case they are blue. Then at time *t* we have, for each evidence statement asserting that a given emerald is green, a parallel evidence statement that that emerald is grue. And the statements that emerald *a* is grue, that emerald *b* is grue, and so on, will each confirm the general hypothesis that all emeralds are grue." (*Fact, Fiction and Forecast* , pp. 74-75; fourth edition, with forward by me, Harvard 1983).

6 More precisely, the choice is arbitrary in what Reichenbach called "primitive knowledge" (his term for the stage at which one begins, when one does not have any higher level inductions about inductions to guide one); in "advanced knowledge" one may, of course, look back on the course of inquiry so far and make inductive inferences to decide what it is best to do next.

7 Cf. Peirce, "On Selecting Hypotheses", in *Collected Papers of Charles Sanders Peirce* , vol. V, *Pragmatism and Pragmaticism* , pp. 413-422, Belknap Press, Harvard, 1965.

8 The problem is that even if the mathematical function f is *not* calculable (in the sense in which recursion theorists use the term), we may still be able to prove that it has the value $f(n) = m$ for some particular n and m ; indeed, there are many non-recursive functions *some* of whose values we know. And if we examine the n th group of dots and find the number is not m , we will certainly have "tested" the hypothesis that the number in the group is always given by the function f , even though the function f is not calculable. Incidentally, there is no way of saying in advance which values of a non-recursive function can be *proved* to be values, because the notion of a "proof" is an open-ended one. (No one formal system can fully capture the notion of a mathematical "proof", in consequence of Gödel's Theorem.)

9 Cf. *On Certainty* , §559 "You must bear in mind that the language-game is so to say something unpredictable. I mean: it is not based on grounds. It is not reasonable (*vernünftig*) or unreasonable. It is there---like our life." Note also §499 "I might put it like this: the 'law of induction' can no more be *grounded* than certain particular propositions concerning the material of experience."

TABLE OF CONTENTS

CHAPTER I

THE GENERAL NATURE OF THE PROBLEM
AND OF THE REQUIRED SOLUTION

Introduction

The problem of which we are treating attains its significance as a result of its connection with another prob-
lem---that of the 'justification of induction'. It is
with this latter problem, accordingly, that we begin.

The problem of justifying induction is one form of
the central problem of the Theory of Knowledge. This is
often thought to be expressed by the question 'How is
knowledge possible?' or 'What are the presuppositions of
knowledge?'[1]. The sceptic insists that a prior question
has been overlooked in the tacit assumption that we do in
fact possess knowledge. Thus epistemological controversy
begins, from Greek times, with the question: 'Have we any
knowledge at all?'.

Now then, doubt as to our possession of knowledge may
arise in two ways. One may be sceptical as to the reliability of the senses---this is the sort of doubt that was
most important in ancient scepticism with regard to knowledge[2]---or dubious as to the possibility of going in any
way beyond the raw data of immediate experience by inference. It is this second ground for doubt that concerns us.

Inferences, again, may be divided into two types, the inductive and the deductive. Deductive inferences---the syllogism is the classic model---are characterized by the property that the conclusion tells us no more than is stated in the premises, as Bacon[3] pointed out (the sense in which this is so will become clearer below). The Schoolmen held that in an analytic proposition the predicate is contained in the conception of the subject. It is the thesis of empiricism that all deductive inferences are analytically valid[4]: that is, for every valid deductive inference we can find a corresponding analytic proposition (namely, the implication whose antecedant is the conjunct of the premises of the inference and whose consequent is the conclusion of the inference in question).

Accepting this point of view, which means granting that the conclusion of a deductive inference in some sense only makes explicit what is already stated implicitly in the premises, we conclude that by means of deductive inference we can go from premises which concern present experience only to conclusions which concern present experience or to tautologies. But that predictive knowledge of which science essentially consists[5] lies forever beyond our reach. Only inductive inference is capable of giving us this, of opening to us the domain of predictive and probable knowledge.

Without induction, knowledge is thus doomed to remain at best a blow-by-blow account of immediate experience; a diary embellished with tautological transformations. Such a 'knowledge' must leave out all that we know as empirical science. It is for this reason that scepticism as to induction is virtually equivalent to scepticism as to knowledge.

This is not to say that one may not have other sorts of doubts as to the status of what we call our knowledge. Doubts as to the trustworthiness of the senses have already been mentioned, and the discussion of these belongs to another chapter of epistemology, one which today concerns itself with such questions as 'How do we know that there is an external world?'. The Greeks[6] managed even to doubt the validity of deductive inference. But their doubts trouble few, if any, modern philosophers. The empiricist saves himself from such an excess of scrupulosity with regard to his scepticism by holding that the deductively certifiable propositions and the analytic coincide. Thus, the validity of such inferences is as certain for him as 'the great truth that there are exactly three feet in a yard'[7].

But this means of guaranteeing the validity of deductive inference pushes all inferences which lead from past and present data to assertions about the future into

the class of inductive inferences, and likewise all inferences which lead from true premises to conclusions which
can be maintained only as probable (since a defining
characteristic of a deductive inference is that the conclusion must be true is the premises are). Thus the problem of in some sense validating inductive inference becomes only the more crucial and the more difficult.

To David Hume belongs the credit for having first
realized the full seriousness and difficulty of this problem[3]. Consider, he argues, any inference from the past to
the future. If the premises of the inference are simply
and solely reports of past experience, the inference cannot be deductively valid---this is simply to say that we
may be correct in our description of our experience, but
deceived in our expectations. Thus, to be deductively
valid, such an inference must contain as premise some principle warranting our step in drawing a conclusion from
matters of past experience as to the nature of our future
experience.

But how may this principle in turn be proved? It is
neither a matter of immediate experience, nor certifiable
as true by logic alone. It must, therefore, be the conclusion of an inference from facts of experience. But
such an inference, we have just seen, requires the use of
this or some similar principle. Thus we are involved in

either a vicious circle or a regress to infinity. In short: such inferences, inductive inferences, cannot be deductively validated. Inductive inference cannot be deductively justified if 'justified' is to mean 'proven to be valid'... or even 'proven to be valid more often than not'.

Clearly, a justification of induction must be either inductive or deductive, like any other argument. If it is inductive---if we use induction to ultimately prove that induction is justified---then we are again arguing in a circle, and we must agree with Aristotle that this is a poor ground on which to rest all of scientific knowledge. This, he says[9], "reduces to the mere statement that if a thing exists, then it does exist---an easy way of proving anything". But, if Hume is right, then induction cannot be deductively justified either, and hence cannot be justified at all. Thus it is that Hume found himself forced to the conclusion, which Bertrand Russell accepts today[10], that we have no rational warrant for induction---we cannot even show that it is reasonable to expect the sun to rise tomorrow![11]

If we are to extricate ourselves from this unfortunate position, two requirements must be fulfilled, as may be seen from the foregoing discussion: any justification of induction must, first of all, be deductive in nature, and, secondly, must justify induction in some sense other than that of proving validity.

To introduce an analogy which we shall find useful in the course of our investigation: we may think of the events in which we are interested as constituting a game. Our 'moves' in this game are our predictions. If we had a proof that induction is valid, we should have no problems in this game, for we would have a winning strategy---that is, a technique for making predictions that we could prove to be good. In the absence of such a proof that induction is a winning strategy, we may at least ask for a proof that it is the <u>optimum strategy</u>[12]---that is, that it is a strategy which will succeed if any strategy will. Such a justification would amount to showing that induction is, in a sense, the 'best bet'.

Such a justification has been given by Reichenbach[13]. What Reichenbach has done---the details we shall for the moment postpone---is to formulate a rule of induction, and to show that if certain conditions are fulfilled (and it can be demonstrated that if these conditions are not fulfilled, then no method, however far-fetched, can lead to successful prediction) the continued application of this rule will lead to (or asymptotically approach) correct prediction. If successful prediction is possible, the continued application of Reichenbach's rule will eventually lead to it.

The problem with which we shall be concerned in the body of this work is partly logical and partly epistemological in nature. The purely logical part has to do with the construction of an interpretation of the concept 'probability' for finite sequences which shall satisfy certain axioms and formal properties, and at the same time permit us to carry through an argument similar to Reichenbach's showing that the use of the Rule of Induction is justified. At the moment it is with the philosophical, epistemological significance of the problem that we deal.

As Reichenbach has pointed out[14], the term 'eventually' in the statement that (if the necessary conditions are fulfilled) by continued application of the Rule of Induction we shall eventually arive at successful prediction, covers a multitude of sins, or at any rate, a multitude of possibilities. It is, for example, perfectly possible that continued application of the Rule of Induction will lead to success, but only after an interval longer than the life-span of the whole human race. In such a case it is clear that the Rule of Induction would be a useless one for human beings in search of success in prediction. Thus, if Reichenbach's justification is to be any justification at all, it is necessary to show something more. It is not enough to say that the Rule of Induction will lead to success if any method will; we must be able to show

that the Rule of Induction will lead to success within the
period in which we are interested in arriving at good predictions if any method will do this. Only then can we say
that the use of the method of induction represents our
'optimum strategy'.

Reichenbach's justification would be a sufficient one
(without this extension) for an immortal observer. For
such a 'player' would be willing to adopt a method, even
though it involved him in error for a finite time, however
long, provided it led to correct prediction for the whole
infinite remainder of his 'play'. Peirce, who believed
that the adoption of inductive methods as a guide to action
involves an emotional identification with an eternal on-
going community of investigators[15], would thus have good
ground to be satisfied. But we are interested in playing
our (hypothetical) game for only a finite time. We propose
to show, by carrying out a program suggested by Reichenbach
who has suggested that the use of infinite sequences in the
theory of probability and in the justification of induction
is merely an idealization, convenient for mathematical
reasons but eliminable in principle[16], that in this case to
it is possible to demonstrate that the use of induction is
our 'best bet'.

In order to carry out this task, it will now be necessary to finitize[17] the theory of probability. Reichenbach's

justification is based upon a theory of probability in which probability is defined in terms of infinite sequences. This procedure is mathematically the most convenient to follow, and for this reason it is customary to overlook the fact that the sequences of events with which human beings deal are finite, though very long, and to treat them as infinite sequences. This assumption is equivalent to the assumption that the 'game' we are playing is an infinite one, and accounts for the peculiarity we have observed in Reichenbach's justification. In order to extend Reichenbach's justification to the finite case, it is necessary to adapt many of the concepts used in connection with the theory of infinite sequences---and in particular the concept of probability---to finite sequences.

The task is somewhat technical in nature. Its significance, however, is broad. For unless we can carry it out, we cannot refute the contention of Hume---the contention of Russell, today---that we have no rational ground to rely upon induction.

Probability and Induction

There is, historically as well as logically, a close connection between the topics of probability and induction. Hume recognized that, while the conclusion of a deductive argument follows of necessity from its premises, the same

cannot be said of an inductive argument. What we know by
means of deductive reasoning alone, we know with certainty.
What demands induction for its establishment, can at best
be said to be known with probability, and thus the theory
of induction becomes the theory of probable knowledge.

While Hume hoped to found such a theory of probable
knowledge[18], he found himself forced to the sceptical conclusions of which we have spoken above. And succeeding
discussion for a long time seemed to show that connecting
the problem of induction with the analysis of the concept
of probability only worsened rather than clarified the
difficulties of the problem.

In the meantime, mathematicians, untroubled by the
philosophic difficulties, were busily constructing a
mathematical theory of probability. One attempt to find a
philosophical ground for this calculus was Laplace's
famous Principle of Indifference[19]. While this principle
has mathematical difficulties, and even contradictions,
which have been pointed out by many writers[20], its main
defect from a philosophic standpoint is that it violates
the fundamental tenet of empiricism in holding nature to be
governed by an apriori principle.

Attempts were also made to find a solution to these
difficulties from an empiricist[21] standpoint (from a ration-
alist standpoint, there is hardly a problem here---only a

choice as to which principle shall be accepted as a priori.)
Some writers sought to cut the Gordian knot, either by so
defining probability as to make it analytic that the conclusion of an inductive inference is probable[22]; or by so
defining it as to make the principle of indifference analytic[23]

These 'solutions' only replace the original problem by
another, even more hopeless. We now have no difficulty in
establishing that, for instance, it is 'probable' that the
sun will rise tomorrow---but does this mean that we have any
rational ground to expect it to rise?

Thus we are apparently impaled on the horns of a di-
lemma. If we so define probability that we can show that it
is reasonable to expect what is probable to happen, then we
are in difficulties when we seek to prove that it is reasonable to use induction to determine what is probable; while
if we so define probability that we can prove that induction (or even apriori reasoning) can be relied on to determine probabilities,.then we cannot show that there is
any rational ground to identify the 'probable' and 'what we
ought to expect to happen'.

The Frequency Interpretation of Probability

As we remarked above, while there is disagreement
about the meaning of the concept of probability, there is

agreement about the formal properties of the concept, as
represented by the mathematical theory of probability. To
give a trivial example, a condition which must be fulfilled
by any explication of the concept is that the probability
of a disjunction must not be less than the probability of
either component---"A or B" is not less probable than A
(if 'or' is used inclusively).

By an interpretation of the concept of probability, we
shall mean a precisely defined expression which can be put
for 'probability' in all expressions of the form: 'the
probability from A to B is p' (where p is a real number
between 0 and 1). The expression so obtained is regarded as
giving the meaning (in precise terms) of the original. In
other words, we wish to have an analysis of the meaning of
'probability' in such statements as: 'The probability of
lightning in a rainstorm is 1/2' or 'The probability of
this theory on the basis of such evidence is .98'.

By a 'formal' system---to make use of a term introduced
by Hilbert[24]---is meant a logical system, or axiomatized and
formalized theory, in which only the logical terms are
given an interpretation. Thus, in Hilbert's formal
geometry, the terms 'plane', 'line', etc., which appear in
the axioms and theorems, do not have any interpretation
specified, but the reader is allowed to supply any interpretation which satisfies the system (that is, which makes

the axioms true statements when the terms mentioned are replaced by their definitions under the interpretation). Such an interpretation may be called an admissible interpretation. The axioms of a formal theory do not specify a unique interpretation; but they determine a class of admissible interpretations, all of which are isomorphic as far as purely formal properties are concerned (provided the system is complete).

In the case of probability also, it is possible to set up the mathematical theory as a formal calculus. This has been done by Reichenbach[25], and we shall extend the formalization in the next chapter of this work. We may thus speak of admissible interpretations of the concept of probability. For example, Carnap's c*, the purely mathematical[26] measure-theoretic interpretation[27], and the frequency interpretation (to mention only a few), are all admissible interpretations of the concept of probability. In the next chapter we shall present another interpretation, related to the frequency interpretation, but dealing with finite sequences, which will be yet another admissible interpretation of the concept. From the plurality of admissible interpretations, it will be seen that the requirement that any interpretation must be 'admissible' is not a very serious restriction; but it is the first condition of which we have to take account.

In order to show that an interpretation is an admissi-

ble one, it is necessary to show that the axioms of the
system can be derived from the interpretation (that is,
that they become analytic when so interpreted). The first
modern investigator to derive extensively the mathematical
theory of probability from the frequency interpretation was
Von Mises[28], who, however, does not draw a sharp distinction
between the interpreted and the uninterpreted system.
Reichenbach, who, as remarked above, was the first to present the mathematical theory of probability as a formal system in the sense described, gives a derivation of the axioms of his system from the frequency interpretation[29], thus
proving the admissibility of this interpretation.

In order to explain in detail the nature of the frequency interpretation, let us first examine a little more
closely the logical structure of the probability statement.
Consider, for example, the statement: 'If this die is
thrown, the probability that it will land on the table
with face one uppermost is 1/6'. Let the sequence

$$x_1, x_2, x_3 \ldots$$

be the sequence of throws of the die, and let the sequence

$$y_1, y_2, y_3 \ldots$$

be the sequence of events of the die coming to rest on the
table after a throw. (We suppose the subscripts to corres-
pond; that is, the event y_{10} is the die coming to rest on
the table after the throw x_{10}, etc.) Then the quoted

statement tells us: 'After any throw x_1, the probability that y_1 is an ace is 1/6'. If we let A be the class of throws of the die, and B be the class of events in which the die lands on the table with face one uppermost, the statement may be phrased (using a variable subscript):

1) For any i, if x_1 is an A, then with probability 1/6, y_1 is a B.

The clause 'x_1 is anA' happens to be superfluous in this example, since the class A includes all the x_1. Such a refernce class (the class A in statements of the form 1) is called the 'reference class', and the class B the 'attribute class') is known as a 'compact reference class'[30]; but not all reference classes need be compact.

With the usual logical notation, plus the symbol '↩' introduced by Reichenbach[31], the statement 1) may finally be written:

2)   (i) $(x_1 \, e \, A \, \underset{1/6}{\leftharpoonup} \, y_1 \, e \, B)$

Statements of the form 2) will also be abbreviated:

3) $(A \, \underset{p}{\leftharpoonup} \, B)$[32]

The form 2) represents the form of a simple, or atomic, probability statement as it appears in the formal calculus of probability. The only extra-logical sign used in such statements is the symbol of 'probability implication', '↩'.

The task of interpreting the probability calculus accordingly reduces to the problem of assigning a meaning to this connective: different interpretations may be distinguished by the different definitions they give it[33].

Let us suppose, now, that we have a finite sequence, say $x_1, x_2, x_3 \ldots, x_{10}$, in which each member is to be regarded simply from the standpoint of its possessing or not possessing a certain attribute A. If a member possesses the attribute we put an A underneath it; otherwise we put $\bar{A}$ (non-A). Thus:

4) $x_1, x_2, x_3, x_4, x_5, x_6, x_7, x_8, x_9, x_{10}$
        A A A $\bar{A}$ $\bar{A}$ A $\bar{A}$ A $\bar{A}$ $\bar{A}$

By the relative frequency of A's in the sequence is meant simply the number of A's divided by the total length of the sequence (in this case, ten). In the example given, the relative frequency of A's is evidently 1/2.

Let us further suppose that we are given a second sequence, $y_1 \ldots, y_{10}$, which is to be similarly regarded from the standpoint of a second attribute B:

5) $y_1, y_2, y_3, y_4, y_5, y_6, y_7, y_8, y_9, y_{10}$
        $\bar{B}$ B B $\bar{B}$ B $\bar{B}$ $\bar{B}$ B B B

By the relative frequency from A to B in the two sequences (in symbols: '$F(x_1 \ e \ A, \ y_1 \ e \ B)$') we mean the number of y_1 that possess the attribute B when the corresponding x_1 possesses the attribute A divided by the total number of A's. If we let the symbol '$N(x_1 \in A.y_1 \in B)$' mean the number of values for which x_1 is an A (that is, the number of A's), then we may put:

6) $F(x_1 \in A, y_1 \in B) = \dfrac{N(x_1 \in A.y_1 \in B)}{N(x_1 \in A)}$

In the example given, the relative frequency from A to B is 3/5. This means that, as we run through the two sequences, when a member of the first sequence is an A, three-fifths of the time the corresponding member of the second sequence will be a B. To put it another way: if every time that x_1 is an A we bet that y_1 is a B we shall win our bets three-fifths of the time.

If the sequences $x_1, x_2, \ldots$ and $y_1, y_2, \ldots$ are infinite, then of course it will be meaningless to speak of the relative frequency from A to B (since in general both numerator and denominator of this fraction will be infinite). But we may speak of the relative frequency from A to B in the first n terms of the two sequences (in symbols: '$Fn(x_1 \in A, y_1 \in B)$'; or to abbreviate as we shall henceforth: '$Fn(A,B)$'). Thus $F_{10}(A,B)$ is the relative frequency from A to B in the two finite sequences obtained by taking only the first ten terms of the x-sequence and the y-sequence respectively.

If we form the fraction Fn(A,B) for larger and larger values of n, it may be that the sequence of numbers so obtained approaches a definite limit p. That is, we may find that for very large n, Fn(A,B) differs in value from p very slightly and can in fact be made to differ from p by as little as we wish provided we take n sufficiently large. In this case, we call p the 'limit of the relative frequency from A to B' (in symbols: $'p = \underset{n \to \infty}{L}\ Fn(A,B)'$).

More precisely: we say that p is the limit of the relative frequency from A to B if and only if for every k, however small, it is possible to find a positive integer N such that

$$\left| Fn(A,B) - p \right| < k$$

whenever n ⩾ N.

For example, the meaning of the statement that the limit of the relative frequency from A to B is 3/5, is that in the long run when x_i is an A, y_i will be a B about 3/5 of the time. More exactly, if whenever x_i is an A I wager that y_i will be a B, the ratio of wins to total bets will in the long run approach more and more closely three-fifths. Similarly, if I say that the limit of the relative frequency of aces in the sequence of throws of a die is 1/6, I mean that in the long run the fraction of the total number of throws in which the die will land on the table with face one uppermost will more and more closely approximate 1/6 as more throws are made.

In terms of the concept of limit of the relative frequency, it is now easy to give the frequency interpretation of probability statements---an interpretation that may be considered to be implicit in Aristotle's statement that 'the probable is what most often happens'[34]. According to this interpretation, to say that the probability from A to B is p means simply that p is the limit of the relative frequency from A to B. Thus the statement 2) has the same meaning as

7) $L \underset{n \to \infty}{} Fn(A, B) = 1/6$

We remarked above that any interpretation of the concept of probability faces two questions: Why is it rational to expect an event which is 'probable' under the interpretation to occur? And how can probabilities be empirically determined? The second question will be dealt with below in connection with the Rule of Induction and the justification of induction. The first admits of an easy, indeed an immediate, solution under the frequency interpretation.

Suppose we know that under certain circumstances, A, an event B will probably occur---say, with a probability of .90. Why is it a good policy to expect B to occur if A does? Because, so the answer of the frequency theorist runs, to say that the probability from A to B is .90 means

that the limit of the relative frequency from A to B is .90
---that is, if whenever A happens I wager that B will happen, in the long run I will be right 90% of the time. In
the absence of a better, it is a reasonable policy (when A
occurs) to bet that B will occur, since we wish to adopt
that policy in our making of predictions which will most
often be successful, or at the least, to be successful more
often than not.

The Rule of Induction

We have stated above that it is possible to formulate
a rule of induction for which (in the infinite case) it is
possible to show that repeated application will lead to
success if success is attainable at all. That is, if the
preconditions for successful prediction are satisfied, the
continued use of the rule will lead to at most a finite
number of errors, and to correct statements (within any
predetermined degree of approximation) for the whole infinite remainder of its employment. We now ask if there is
not more than one such rule.

Evidently, any two rules, both of which have this property, must agree on all the predictions they yield (within the interval of approximation) after a certain point.
That is to say that if the values obtained from the first
rule are $a_1, a_2, a_3 \ldots$, the values obtained from the second

rule must be $a_1 \neq f(1)$, $a_2 \neq f(2)$,..., $a_i \neq f(i)$,...
where $f(n)$ is a function which converges to 0 as n becomes larger[35].

Let the rule A and the rule B be any two rules for
which we can give a deductive justification of the type in
question. Then, by the argument just given, we can show
that the rule B has the form: apply rule A and add $f(i)$,
where i is the number of the application in question, and
$f(n)$ is some function that goesto zero as n becomes infinite.

Conversely, let A be the rule for which we have already given the justification, and let B be any rule of the form:
apply A and add $f(i)$ where $f(n)$ is _any_ function that becomes
zero as n becomes infinite. Then after a certain point the
values given by rule B will be as close to the values given
by rule A as we like, and these in turn, by our justification,
lie as close to the true values as we like; after a certain
point rule B gives correct results within any predetermined
interval of approximation[36]. Thus the rule B is also justified.

A formal proof of the statement just made runs as follows: the statement that after a certain point the A values,
that is the a_i, are all correct within the predetermined interval of approximation means, in more precise terms, that
the difference $|a_i - t_i|$ (where t_i is the true value) conver-

ges to zero as i becomes larger. Furthermore, we know that
the difference $|a_i-b_i|$ also converges to zero. Now then, let
us choose an interval of approximation d $>$ 0. Then, from
the definition of convergence[37], it follows that for some in-
teger N_1 it is true that whenever i$>$ N1, $|a_i-t_i|<$ 1/2d, and
for some integer N_2 it is true that whenever i$>$ N_2, $|a_i-b_i|$
$<$ 1/2 d. Hence if K is the larger of N_1 and N_2, we have
that whenever i$>$K, both $|a_i-b_i|<$ 1/2 d and $|a_i-t_i|<$ 1/2 d,
and hence $|b_i-t_i|<$ d. But this simply means thatthe differ-
ences $|b_i-t_i|$ converge to zero as i grows without bounds.
Hence, whatever value we may have selected for the interval
of approximation d, after a certain point K the B values
will all be correct (will differ from the true values by less
than d). Hence the B method is justified.

Let us recapitulate: we have shown that if a rule A
is justified, any other justified rule B must have the form
A $\neq$ f, where f is a function that converges to zero, and,
conversely, any rule of the form A $\neq$ f is justified. Thus,
if we can determine one justified rule A, we shall have determined the whole class of rules that can be justified in
this sense.

Before stating such a rule, let us first consider what
we mean by 'prediction'. We have already discussed, in the
conclusion of the foregoing section, how our knowledge of
limits of the frequency controls our predictions of indivi-

ual events. It is true that there are some problems that
e have not discussed in this connection: in particular the
roblem of choosing the sequence and the reference class to
hich we shall regard as belonging the individual event that
e are interested in predicting. This problem is one that
an itself be reduced to a determination of the limit of an
nfinite sequence, however. For let us consider two alter-
ative methods of so classifying individual events in order
o determine the statement we should make as to the outcome:
vidently, the better method is the one that leads to the
reater proportion of true statements in the long run, in
hort, the one with the higher 'success-ratio'. The deter-
ination of the success-ratio that will be obtained if we
ollow a given inductive method is simply a determination
f the limit of an infinite sequence---one whose members
are themselves inductive inferences---and the problem of
etermining which of a set of inductive methods has the
ighest success-ratio is a problem in the theory of induc-
ion on a higher level ('cross-induction'[38]). What is important for us is the logical form of this problem: the
etermination of the limit of an infinite sequence, or of
a set of infinite sequences.

For this reason, the interpretation that we give to
the scientific problem of good prediction is the correct
etermination of limits of the relative frequency (still
aking the assumption that all our sequences are infinite).

By a 'prediction', therefore, we mean a statement as to the relative frequency with regard to an infinite sequence or pair of infinite sequences[39]. By the 'true value' we mean the actual limit of the frequency: and by the justification of a method we mean a demonstration that the values obtained by repeated applications of the method converge to the true limit of the relative frequency.

Accordingly, when we speak of a rule of induction, we mean a rule for making a series of approximations to the limit of the relative frequency (with regard to any particular attributes A and B, and any two infinite sequences). We shall now state such a rule, leaving the question of its justification to the following section.

This rule, as given by Reichenbach[40], formalizes the simplest of the classical modes of inductive inference---, induction 'by simple enumeration'.[41] It is simply to 'posit' (or wager) that the observed relative frequency will persist. That is, if the relative frequency in the portion of the sequence that we have so far examined is k, we predict that the limit of the relative frequency is in the interval $k \not< d$.

The term''posit' requires some explanation. A posit, as the term is used by Reichenbach[42], is a statement that is made not because we can show it to be true, but because we can show that the policy according to which it is made is a justifiable one. This is easily made clear by employing the game-theoretic concepts we have so far introduced. A 'posit', in these terms, is simply a 'play' or 'move' that we make as a part of a strategy that we know to be a winning, or at least an optimum, one.

Thus, at the end of the preceding section, we argued that it was a good policy, when A occurs, to predict that B will occur (assuming that we know the limit of the relative frequency from A to B to be .90). This assertion is a 'posit', since we do not know in fact that B will occur, but we make the statement according to a strategy (stating, whenever A occurs, that B will occur) which we know we can expect to succeed in the long run 90% of the time. Such a posit is called by Reichenbach an 'appraised posit'[43], because we have a measure of the goodness of the posit in our knowledge of the frequency with which the strategy according to which it is made will succeed (that is, the frequency with which statements made according to it will be true--- in this example, .90).

Reichenbach also introduces the term 'anticipative posit'. An anticipative posit is, in the first place, not appraised. That is to say: we do not know what the frequency of success of the policy according to which we make a series of anticipative posits will be.

As anticipative posits, we may cite the series of statements that we make when we repeatedly employ the rule of induction. These statements we do not know to be true. We do

not even know that the law contains more truths than falsehoods. Nevertheless, the strategy we are following in making these statements is a good one, because we know that it
will lead to truth if any method will do this. That is to
say, we know that it is an optimum strategy.

Thus the difference, in Reichenbach's terminology, between 'appraised' and 'anticipative' posits, corresponds to
the difference between a winning and an optimum strategy,
in our terms. An 'appraised' posit is a statement made according to a winning strategy; and the 'appraisal' of a posit is simply the success-ratio of the strategy. An 'anticipative' posit is a statement made according to an optimum
strategy: we do not have a knowledge of the frequency of
true statements in a sequence of anticipative posits simply because we do not know that the strategy we are following will win (lead to a majority of truths); we only know
that it will do so if the conditions for success exist.

We may now state precisely the Rule of Induction:

If we have examined the first n terms in a pair of sequences, and in this finite initial section the relative frequency $F_n(A, B)$ is k, posit that the relative frequency $F_i(A, B)$
approaches a limit in the interval k∠d as i increases without limit.

The Justification of Induction

The aim of science, we have remarked, is prediction, by which we understand the determination of the limits of infinite sequences. To justify induction, accordingly, means to show that some rule of induction will attain this aim if it can be attained; that it will lead to a correct statement of the limits of infinite sequences if anything will---that is, if there are, among the sequences of events in which we are interested, any that possess a limit.

We may also employ in this connection an argument that has been mentioned above. To say that there is a good method for making predictions is to say, at the very least, t. that there is a method which, when employed to make statements about the outcome of future events, leads in the long run to a balance of truths over falsehoods. That is, in the infinite sequence of statements made according to this method, there is a high success-ratio. But a 'success-ratio', or a 'long-run frequency of successes' in an infinite sequence, is nothing but the limit of a relative frequency. The precondition for successful prediction is the existence of sequences which possess a limit of the relative frequency.

Our aim, then, is the determination of limits; and our procedure in justification is to show that if limits exist, our rule will find them. Reichenbach has compared this type of justification with the arguments employed in several familiar situations to prove the reasonableness of certain lines of action[44]: For example, Magellan's policy in sailing along the coast was a reasonable one, not, indeed, because he knew that he would find a passage to the East, but because he knew that he would find one if one existed. In the absence of any knowledge that there is not such a passage, this constitutes a sufficient justification of such a plan. Similarly, if we wish to catch fish in a certain lake, not knowing whether or not there are any there, it is a reasonable policy to throw in a net because it will enable us to catch them if they canbe caught at all On the same grounds, Reichenbach argues[45], it is a reasonable policy to use the rule of induction: our aim is to determine limits; we do not know whether or not this aim can be achieved; but we employ the rule because it will determine limits if they can be determined.

This type of justification has been spoken of above[46]. In our terms, what Reichenbach shows is simply this: that i the 'game' where our predictions are our 'moves', the use of the Rule of Induction is an optimum strategy. To show this involves showing that the series of values obtained from the rule will eventu ally reach, and remain at, the true limit of the frequency (within the degree of approximation chosen, whatever it may be). But this is simply to show that the values given by the rule converge to the limi of the frequency[47]. But the values given by the rule are the finite relative frequencies $F_n(A,B)$, $n=1,2,...$; and the limit of the relative frequency is by definition the number to which these frequencies converge. Thus, if our sequence possesses a limit, the rule gives values which converge to it: briefly, the use of the rule leads to an eventual determination of the limit.

Two objections may be raised against this justification (in addition to the one which it is the business of this work to analyze): that it is not enough to know the limit of the frequency in those cases where our sequence has a limit of the frequency---we also wish to know the sequences which do not possess a limit; and that it may be that there is a method that will determine the limit more efficiently than will the Rule of Induction.

To refute these objections it is necessary to prove two things : that the Rule of Induction can be used todetermine when a sequence does not possess a limit; and that it can be used to find a better method if one exist. The proof of these statements we leave for a later section.

The Problem of the Finite Case

We can now give a complete statement of our problem: it is to produce a precise interpretation of probability which 1) defines the probability of an event in a finite sequence of events, or better, the probability that such an event will have a property B if it (or a cooresponding event in another sequence) has the property A; 2) is admissible, that is, satisfies the axioms of the calculus of probability; and 3) is justifiable, that is, makes it possible for us to show that the rule of induction is the best instrument for the determination of probabilities so defined.

The reason that we have stated the first requirement in this form lies in the logical structure of the probability statement; this, and the second requirement, have been discussed above under the frequency interpretation. The kind of justification that we are seeking has to some extent been indicated in the Introduction;how this kind of justification can be given (a proof that the use of the Rule of Induction represents the best strategy) will become clearer in the body of this work.

To recapitulate: we regard human beings as being interested in certain finite sequences of events from the standpoint of predicting what the nature of those events will be. We wish to explain what we mean, when we say that it is 'probable' that under certain conditions one of those events will have a certain attribute, in such a way that it can be shown that the 'high probability' of an event is a rational ground for positing that it will occur, and in such a way that it can be shown that the repeated use of the Rule of Induction is the best instrument for determining what the probability of an event is.

Our concern at the moment is not with the solution
to this problem, but with laying down the requirements that
any solution must conform to in order to constitute a solution.   These requirements, as we have indicated, are fini-
tistic character, 'admissibility', and justifiability.   Under the last of these requirements we understand both the
justification of the use of probabilities as a guide to action and the justification of induction as a method of evaluating probabilities.

CHAPTER II

THE CALCULUS OF PROBABILITY
AND ITS INTERPRETATION

Introduction

The present chapter is largely logical and mathematical, rather than philosophical, in character. Our task is to present the calculus of probability as formalized by Reichenbach, with certain additions and modifications of our own; to present the finite frequency interpretation of the calculus (the interpretation of probability as the relative frequency in a finite sequence), and to show that it is an admissible interpretation; and to show that our choice of this interpretation is justified.

While this task might seem, at first glance, to present merely a forbidding mass of detail, we shall discover, on the contrary, that it yields very surprising results. These results occur mainly in connection with the theory of the order of probability sequences. This theory is constructed by Reichenbach from two axioms, both of which seem to be satisfied only by infinite sequences. We shall show that it is possible for finite sequences to satisfy the requirements of the theory of order, and this demonstration will not require a change in the axioms, but only a change in their interpretation; that is, the modification of a semantical rule tacitly presupposed in the theory as developed for infinite sequences. (This change will not modify the meaning of the axioms for infinite probability sequences). We shall also show that it is possible to characterize different kinds of order in finite probability sequences, corresponding to the different possibilities encountered in the infinite case. This fact is very important for the development of the formal theory of induction.

But the most surprising result we encounter, perhaps innthe whole finite theory, is connected with our argument for choosing the frequency interpretation. The grounds for our choice are both formal and material in nature. The material grounds do not differ from the corresponding arguments for the adoption of the limit-theory when dealing with infinite probability sequences. These are connected with the use we make of probabilities as a guide to action. But the formal requirement has absolutely no analogue in the infinite theory: We shall show that the formal system itself permits of essentially only one interpretation.

The vast number of essentially different admissible interpretations allowed by most formal systems (including the probability theory for infinite sequences) makes the result extremely startling. Yet we shall show that this uniqueness character does in fact obtain for the calculus of probability: It can be shown <u>within the uninterpreted</u>

system that if there are any finite probability sequences
the probabilities in those sequences must be equal to the
relative frequencies.

Thus we shall do more than merely satisfy our first
requirement:[48] We shall show that we have produced an admissible and justifiable interpretation of probability
for finite probability sequences in essentially the only
way possible.

The Probability Calculus

The Calculus of Probability is a formalized and axiomatized mathematical system. Like all such systems, it
presupposes only the laws of logic (including mathematical analysis) for its deductions. In this respect, it
is exactly comparable to geometry, as formalized by Hilbert[49], or point set topology as formalized by Sierpinski.[50]

The subject matter of the system, naturally enough,
is probability: that is, it is concerned with finding
rules by means of which we may determine the values of
certain probabilities when certain other probabilities
are given. For example: When we infer that, if the
probability of obtaining an ace when we cast a die is
1/6 and the probability of obtaining a five is 1/6,
then the probability of obtaining a five or an ace is
1/3, we are making use of a basic theorem of this calculus.

Thus the general form of the theorems of the system
is this: if certain probabilities have certain values,
then certain other probabilities have certain other values.

In our discussion of the logical form of probability
statements[51], we saw that the statement: 'The probability from A to B is p' reads in the expanded logical notation:

8)   (i) $(x_i \; e \; A \; \underset{p}{\leftarrow} \; y_i \; e \; B)$

    (we remind the reader that the sign '$\leftarrow$' is to be
    read: 'implies with probability p'.)

We shall also make use of two forms of abbreviation:
The first consists of dropping the sequence variables in
8)[52], thus:

9) $(A \; \underset{p}{\leftarrow} \; B)$

while the second is the graphic 'P-notation':

10) $P(A,B) = p$

    (to be read: 'The probability from A to B is p'.)

Then, in the formal notation, the statements of the
Calculus will be all the statements of the form 8), and
all the logical compounds of such statements, that is, all
the statements built up from such statements as ultimate
constituents by the application of the truth-functions

('.'---'and', 'V'---'or', '⊃'---'implies', '≡'---'if
and only if', '‾'---'not',), and quantification
('(x)'---'for allx...').

The foregoing is to be taken as an impressionistic
characterization of the system; a precise definition of
'formula of the Probability Calculus' is given by us below.

In formalizing a mathematical system, one requires a
characterization of the formulae of the system and a
statement of the axioms and rules of the system. We shall
employ the axioms and rules of Reichenbach's system,
which we shall henceforth designate as the system R.

These axioms are, at first blush, surprising in that
they are so simple and so few in number. (This is a characteristic of many formal systems---e.g., topology.)
They include: the principle that the probability of B or
C (we omit the general reference class A for brevity; and
we suppose that B and C are incompatible) is the sum of
the separate probabilities of B and C (that is,
$P(A, B \lor C) = P(A, B) \neq P(A, C)$); the principle that the probability of B and C is obtained by multiplying the probability of B by the probability from B to C (that is,
$P(A, B.C) = P(A, B)P(A.B, C)$); the principle of 'univocality'---if the probability exists at all, one and only one
number is the probability from A to B (except in the trivial case in which the reference class A is empty); and
the principle of 'Normalization'---part of which asserts

that probability (except in the trivial case mentioned) is $\geq$ 0. That it is $\leq$ 1 is not an axiom, for this is derivable from the other assumptions. In addition, the principle of Normalization permits us to go from a general implication

$$(i) \quad (x_i \; e \; A \supset y_i \; e \; B)$$

to a probability implication with probability one:

$$(i) \quad (x_i \; e \; A \underset{1}{\bullet} y_i \; e \; B).$$

These axioms constitute virtually the entire framework of the Elementary Calculus of Probability. (So called because it deals with probability sequences as external wholes, whereas the theory of order speaks of their internal composition.)

With the addition of two equally simple principles required for the development of the theory of order, we have here an adequate axiomatic basis for the whole mathematical theory of probability.

The Formalization of the Calculus of Probability: The Definition of Formula

As we study Reichenbach's formulation of the Probability Calculus, we encounter more and more logically complex probability formulae (involving subscripts indicating the place of an element in an n-dimensional 'lattice' of sequences, subscripts employed in drawing inferences concerning 'phase' probabilities, etc.). We soon see that there are far more general formulae than 8) required to express the form of probability statements, and are thus led to seek a syntactical (formal) definition of 'formula of the Calculus of Probability' which shall hold good for the required cases.

The problem of carrying out this program turns out to be surprisingly complex. The result of our investigation we now present:

1) <u>Kinds of Variables</u>

    a) Class variables: The following infinite alphabet is used for classes or attributes[53]:

        $A, B, C, \ldots, Z, \quad A', B', \ldots, \quad A'', \ldots$

    b) Element variables:

        $x, y, z, w, \quad x', y', \ldots \quad x'', \ldots$

    c) Subscript variables (to be appended as subscripts to element variables, as in '$x_{i,j,k}$'):

        $i, j, k, \quad i', j', k', \ldots$

    d) Integral variables:

    (Constants denoting the various integers, and real numbers in general, are regarded as logical signs and hence part of the apparatus presupposed by the system.)

        $\alpha, \beta, \ldots, \epsilon, \qquad \alpha', \beta', \gamma', \ldots, \alpha'', \ldots$

e) Real number variables:

$$o, p, q, r, \quad o' \ldots \quad o'', \ldots$$

2) <u>Constants</u>

We require all logical constants (e.g., '.', 'v'),
and in addition the single extra-logical constant '←'.

3) <u>Terms</u>

    a) Sequence variables:

        If a_1 is an element variable and $b_1, b_2, \ldots b_n$
are the first n subscript variables, in alphabet-
ical order $a_{b_1, b_2, \ldots b_n}^{54}$ is <u>a sequence variable</u>
<u>of degree n.</u>

    b) Subscript terms:

        If $a_1, a_2, \ldots a_{n-1}$ are constants denoting integers
or integral variables, and $f(x_1, x_2, \ldots, x_n)$ is an
integral function for integral values, and b_1 is
a subscript variable, then $f(b_1, a_1, \ldots, a_{n-1})$ is
<u>a term corresponding to the subscript b_1.</u>
Since f may be a constant function, every integer-
symbol is a term corresponding to every subscript
variable b_1.

4) <u>Definition of Formula</u>

    a) Elementary atomic formula:

        If A_1 is a class variable and b_1 is a sequence var-
iable of degree n then:

        $(b_1 \ e \ A_1)$

is an <u>elementary atomic formula of type n.</u>

  b) Elementary molecular formula:

  If $A_1, A_2, \ldots, A_n$ are elementary atomic formulae of
  degree n, and B_1 is a truth-functional compound of
  $A_1, \ldots, A_n$, then B_1 is an <u>elementary molecular for-</u>
  <u>mula of degree n.</u> (Recursively: if A_1 is an elementary atomic formula of degree n, then it is an
  elementary molecular formula of degree n, and if
  B_1 and B_2 are both elementary molecular formulae
  of degree n, and the 'χ' designates one of the
  truth-functional connectives, $(B_1 \chi B_2)$ is an elementary molecular formula of degree n, and so is
  $\overline{B}$.

  c) Elementary probability implication:

  If $(A_1 \supset B_1)$ is an elementary molecular formula
  of degree n, and A_1 is a conjunction of elementary molecular formulae of degree n; and at least
  one of the terms of the conjunction A_1 is an elementary atomic formula of degree n $(a_1 \, \epsilon \, K_1)$
  where the class variable K_1 does not occur with
  any other sequence variable in A_1, $(K_1$ is simply the reference class); and c_1 is a real variable or a real-number-symbol: (and b_1 is one
  of the subscript variables of $(A_1 \supset B_1)$), then
  $(A_1 \overset{b_1}{\underset{c_1}{\multimap}} B_1)$ is an elementary probability implication of degree n.

d) Probability implication:

If $(A_1 \overset{b_1}{\underset{c_1}{\leftarrow}} B_1)$ is an elementary probability implication of degree n, and $(A_1' \overset{b_1}{\underset{c_1}{\leftarrow}} B_1')$ is obtained
from $(A_1 \overset{b_1}{\underset{c_1}{\leftarrow}} B_1)$ by replacing zero or more of the
subscript variables by terms corresponding to
those variables (not including the one above the
'$\leftarrow$'), then $(A_1' \overset{b_1}{\underset{c_1}{\leftarrow}} B_1')$ is a probability implication of degree n.

e) The formulae of the Probability Calculus:

All probability implications of degree n are formulae of degree n, all truth-functions of formulae
of degree n are formulae of degree n, and all quan-
tifications of formulae of degree n are formulae.

The procedure we have adopted here in introducing a
running subscript over the '$\leftarrow$' is a departure from the
notation of Reichenbach[55]. It is introduced for several
reasons.

In the first place, we are more lenient in our definition of formula than R, inasmuch as we tolerate such
expressions as

11) $'(x_{10} \, e \, A \overset{1}{\underset{1/6}{\leftarrow}} y_{10} \, e \, B)'$

as probability formulae. This is done even though, as we
have remarked, probability statements are meaningful only
when they are assertions about properties of sequences.
That is, from the logical standpoint, the form of a proba-

bility statement is

2) $\qquad$ (i) $(x_1 \, \epsilon \, A \underset{1/6}{\overset{i}{\cdot}} y_1 \, \epsilon \, B)$

---its specification instance, 11) above, has no meaning.

However, when the probability in a given sequence is known to have a value p, we may speak of the probability of each single event in that sequence as p. That is to say, we may endow the expression 11) with meaning by agreeing to regard it as having the same significance as 2). The philosophic ground for doing this is simply that the probability p controlling the sequence determines, as we have pointed out[56], our willingness to posit that the single event we are concerned with will belong to the attribute class. We may thus use p as a measure of the reliability of the assertion that the event will belong to the attribute class. So regarded, p is the 'weight' of an assertion about a single event[57].

But while it is the use of probabilities as 'weights' in speaking of single cases that leads to our adoption of the 'fictitious transfer' of probability to single events, at the moment it is not with the material grounds for the transfer but with its formal convenience that we are concerned. What is important for us is that this procedure (identifying 11) and 2) with respect to meaning) legitimizes the logical operation of specification as applied to subscript variables.

The only difficulty that could arise from this opera-

tion, would arise if the operation of specification made
it unclear to what sequence we regarded the single case
in question as belonging. This would occur with Reich
Reichenbach's notation if specification were allowed, and
is the reason that it is eschewed in the system.

Thus, let us suppose that A is not empty, that is,

12) $(\exists i)(\exists j)(x_{ij} \, e \, A)$

and further:

13) $(j)(i)(x_{ij} \, e \, A \underset{p}{\twoheadrightarrow} y_{ij} \, e \, B)$

14) $(i)(j)(x_{ij} \, e \, A \underset{p'}{\twoheadrightarrow} y_{ij} \, e \, B)$

where $p \not\equiv p'$.

The first statement means that the probability from
A to B (with regard to the 'horizontal' sequences
$x_{1j}, x_{2j}, x_{3j}\ldots$ and $y_{1j}, y_{2j}, y_{3j}\ldots$) is p (whatever value
we may give j).

The second statement asserts the probability from A
to B (with regard to the 'vertical' sequences $x_{11}, x_{12}, \ldots$
and y_{11}, y_{12}) is p' (whatever value we may give i).

If we allowed specification we would immediately obtain:

15) $(x_{1,1} \, e \, A \underset{p}{\twoheadrightarrow} y_{1,1} \, e \, B)$

        and

16) $(x_{1,1} \, e \, A \underset{p'}{\twoheadrightarrow} y_{11} \, e \, B)$

but 15) and 16) violate the principle of Univocality,
since A is not empty, and yet they assert two different
values to a probability[58].

In our notation 13) is written:

$$(j) \ (i) \ x_{ij} \ \varepsilon \ A \ \overset{i}{\underset{p}{\bullet}} \ y_{ij} \ \varepsilon \ B)$$

and 15) becomes:

15') $(x_{1,1} \ \varepsilon \ A \ \overset{i}{\underset{p}{\bullet}} \ y_{1,1} \ \varepsilon \ B)$

(the 'running subscript' is regarded as part of
the sign of probability implications and is accordingly
unaffected by the specification).

Similarly, 16) becomes:

16') $(x_{1,1} \ \varepsilon \ A \ \overset{j}{\underset{p'}{\bullet}} \ y_{1,1} \ \varepsilon \ B)$

and the apparent violation of univocality vanishes.

Thus we see the reason that the use of a 'running subscript' permits us to employ the rule of specification
(the only remaining restriction being that the running
subscript remains unchanged. Since it acts as part of a
constant, this is, however, not a real restriction.) It
does this because it 'labels' the statement about the single case, that is, it tells with respect to what sequence
we are determining the probability.

Still more important is the following consideration.
In R, from 16) we can immediately infer (using 12)) and

the principle of Univocality):

17) $\overline{(i)\ (j)\ (x_{ij}\ e\ A \underset{p}{\leftarrow} y_{ij}\ e\ B)}$

But 13) and 17) clearly express a violation of the commutativity of the all-operator, which accordingly has to be abandoned in R.

Thus, if the notation of R is to be maintained, entirely new logical rules would have to be worked out for the all-operator.

In our notation, 13) can be written

$$(i)\ (j)\ (x_{ij}\ e\ A \overset{i}{\underset{p}{\leftarrow}} y_{ij}\ e\ B)$$

or $\quad (j)\ (i)\ (x_{ij}\ e\ A \overset{i}{\underset{p}{\leftarrow}} y_{ij}\ e\ B)$

and 14) can be written

$$(i)\ (j)\ (x_{ij}\ e\ A \overset{j}{\underset{p}{\leftarrow}} y_{ij}\ e\ B)$$

or $\quad (j)\ (i)\ x_{ij}\ e\ A \overset{j}{\underset{p}{\leftarrow}} y_{ij}\ e\ B)$

Thus, not only does the apparent violation of commutativity vanish: the general applicability of this logical principle is insured.

To recapitulate: In R the all-operator is made to do the work of indicating the 'running subscript' in addition to its usual function of universal quantification. Hence, in a complete formalization of R, entirely new and complicated rules would have to be worked out for it.

In our notation, these two functions are separated,

and hence the laws of logic, including quantification theory, can be used in deduction without alteration or limitation.

There is one case, in our notation, where the 'running subscript' is superfluous, however. This occurs whenever the formula is of degree one, for in that case it follows from our definitions that the running subscript must be '1'. We may therefore introduce the following obvious convention of abbreviation: to drop the running subscript in all formulae of degree one.

The Formalization of the Calculus of Probability:
Axioms

The axioms of the Elementary Calculus of Probability as given by Reichenbach, are as follows:

I. Univocality

$$(p \neq q) \Rightarrow (A \rightarrow B).(A \leftarrow B) = (1)\overline{(x_1 \ e \ A)}$$
$$\quad\quad\quad\quad\quad\quad\quad p \quad\quad q$$

II. Normalization

1. $(A \Rightarrow B) \Rightarrow (A \underset{1}{\leftarrow} B)$

2. $(\exists i) (x_1 \ e \ A).(A \underset{p}{\leftarrow} B) \Rightarrow (p = 0)$

In the statement of these axioms, we have employed an abbreviation, dropping the sequence variables. That is, '(A⊃B)' abbreviates the formal implication '(i) $(x_1 \ e \ A \Rightarrow y_1 \ e \ B)$', and similarly '(A⊃B)' abbreviates the general probability implication '(1)$(x_1 \ e \ A \leftarrow y_1 \ e \ B)$'. Reichenbach has produced a complete formal rule for ab-

breviating formulae of the system[59]. We shall not state
the rule here, (though the foregoing abbreviations represent an application of it) but will continue to abbreviate as we have, in the following axioms.

III.   The Theorem of Addition

$$\vdash (A \underset{p}{\leftarrow} B).(A \underset{q}{\leftarrow} C).(A.B \supset \bar{C}) \supset (\exists r)(A \underset{r}{\leftarrow} BVC).(r\text{-}p/q)$$

IV.   The Theorem of Multiplication

$$\vdash (A \underset{p}{\leftarrow} B).(A.B \underset{u}{\leftarrow} C) \supset (\exists w) (A \underset{w}{\leftarrow} B.C).(w\text{-}p.u)$$

It should be noted that when substitution is employed in these axioms, the same running subscript must
appear above each occurrence of the '←' in the formula
obtained by the substitution.

In addition to these axioms, and the Rule of Existence (of which we speak below), a further principle is
frequently employed in deductions. This pronciple,
which we shall now formulate, expresses the extensionality of the system. That is to say, it expresses the fact
that $(A_1 \underset{c_1}{\leftarrow} B_1)$ is an extensional context of all the
expressions occurring in it (to use Carnap's[60] terminology).
In other words, if an expression occurring in a formula
can be shown to be materially equivalent to another expression, we are justified in substituting this other expression for the first one in one or more of its occurrences in the formula; and the resulting formula will be
equivalent to the original. Formally[61]:

    The Principle of Extensionality

If C_1 is like C_1' except for containing some occurrences of A_1 where C_1' contains occurrences of B_1, then

$$(A_1 \approx B_1) \supset (C_1 \approx C_1')$$

In spite of its resemblance to certain logical principles, it should be observed that this is a principle of the system and not a logical law. For nothing in logic, not in the rest of R, tells us that the '$\approx$' of probability implication is not an intensional symbol (like, for example, Lewis' 'strict implication' '$\dashv$'). It is, furthermore, an extremely important principle; without it, not even the general theorem of addition can be derived.

In addition to its formal importance in all developments of the mathematical theory of probability, the principle has a philosophical significance that should not be overlooked. For it rules out all intensional interpretations of '$\approx$', and hence, in particular, such interpretations as 'Probability is the degree of belief.' (For such statements as: 'If A occurs, then x will expect B with intensity p' are intensional, like all belief sentences.)[62]

The Formalization of the Calculus of Probability:
The Rule of Existence

In addition to the axioms so far enumerated, R contains an important rule called the Rule of Existence.

As stated by Reichenbach[63] the rule is:
"If the numerical value p of a probability implication $(A \underset{p}{\rightarrow} B)$, provided the probability implication exists, is determined by given probability implications according to the rules of the calculus, then this probability implication $(A \underset{p}{\rightarrow} B)$ exists."

This rule may be interpreted in two ways. If we take it literally, we must understand it to mean the following (where $(P_1 \underset{c_1}{\rightarrow} Q_1), \ldots, (P_n \underset{c_n}{\rightarrow} Q_n)$ and $(A_1 \underset{d}{\rightarrow} B_1)$ are probability implications of the same degree, and with the same reference class, and where 'f' designates the name of an analytic function with n variables):

The Simple Rule of Existence

a) If $\vdash (P_1 \underset{c_1}{\rightarrow} Q_1) \ldots (P_n \underset{c_n}{\rightarrow} Q_n) . (\exists d)(A_1 \underset{d}{\rightarrow} B_1) \supset d$
$= f(c_1 \ldots c_n)$

then $\vdash (P_1 \underset{c_1}{\rightarrow} Q_1) \ldots (P_n \underset{c_n}{\rightarrow} Q_n) \supset (\exists d)(A_1 \underset{d}{\rightarrow} B_1)$

This expresses what we may call a single application of the Rule of Existence. In such an application, we assume the existence of a single probability in addition to the given probabilities; we show that this probability is uniquely determined by the given values; and we conclude that the probability does in fact exist.

Many applications of the Rule of Existence are not of the same sort, however, but rather of the following sort: Assume that a set of probabilities, $d_1, d_2, \ldots, d_n$, all exist. If we are able to show, with this assumption, that each of the d_i is a function of the given probabilities, then (without the assumption) we may conclude that all

the d_i exist. Such an application is made, for example,
when we assume that three probabilities exist: Infer that
we can derive three simultaneous equations in these probabilities with functions of the given probabilities as
coefficients; solve the system to express each of the
three as a function of the given probabilities; and finally infer, with the Rule of Existence, that all the
three. exist.

The fact that uses of this kind are made of the Rule
of Existence[64]might lead us to conclude that the rule
would be better expressed as follows:

The Simultaneous Rule of Existence

b) If:

$\vdash (P_{1c_1}Q_1).(P_{2c_2}Q_2)...(P_{nc_n}Q_n).(\exists d_1)(A_{1d_1}B_1)...(\exists d_m)(A_{md_m}B_m) \supset$

$d_1 = f_1(c_1,...c_n).d_2 = f_2(c_1,...,c_n).d_m = f_m(c_1,...,c_n)$

Then:

$\vdash (P_{1c_1}Q_1)...(P_{nc_n}Q_n) \supset (\exists d_1)(A_1 \, d_1 B_1)...(\exists d_m)(A_m \, d_m B_m)$

Which form of the rule shall we adopt?

This problem bears a strong formal analogy to the
choice of a rule of substitution in the propositional
calculus. On the one hand we may choose the <u>Simple Rule
of Substitution</u>:

c) If S_2 is obtained from S_1 by putting a formula A
for all occurrences of a variable p in S_1; then S_2 may
be inferred from S_1.

On the other hand, we have the <u>Simultaneous Rule of Substitution</u>:

d) If S_2 is obtained from S_1 by putting formulae $A_1, A_2, \ldots, A_n$ for all occurrences of $p_1, p_2, \ldots, p_n$ respectively,[65] then S_2 may be inferred from S_1.

The difference between c) and d) is strikingly analogous to that between a) and b). a) and c) both have the form: 'If a single quantity has a certain property, then...', while b) and d) are of the form: 'If a <u>set</u> of quantities simultaneously have the property, then...'. The problem of a choice between c) and d) in the propositional calculus is obviated by showing that they do not differ in real strength: Any result that can be obtained by the use of the apparently stronger rule d) can be obtained by using c) alone.

We shall solve the problem that confronts us in a similar way: We shall show that a) is a sufficient interpretation of the Rule of Existence, by showing that any result obtainable from b) is obtainable by the use of a) alone. In a certain sense, this amounts to deriving b) from a) (since, if we can give such a proof, we can assert b) as a metatheorem).

In stating this proof, we shall abbreviate $(P_{1c_1} \dashv Q_1)$ by K_1, $(A_1 \underset{d_1}{\mathbf{f}} B_1)$ by L_1, and $f_1(c_1, c_2, \ldots, c_n)$ by f_1.

$$\bar{z} \cdots \bar{a} \cdots 3$$

Theorem:

If $\vdash K_1 . K_2 ... K_n . (\exists d_1) L_1 ... (\exists d_m) L_m \supset$

$d_1 = f_1 . d_2 = f_2 ... d_m = f_m$, and all the d_i are distinct from one another and from the c_i^{66}, then

$\vdash K_1 . K_2 ... K_n \supset (\exists d_1) L_1 . (\exists d_2) L_2 ... (\exists d_m) L_m$

(where we allow ourselves to use in derivation only the <u>Simple</u> Rule of Existence.)

Our proof will be an inductive one. We suppose that the theorem has been proved for $m = M - 1$, and show that it follows for $m = M$. (For $m = 1$, the proof is immediate since the theorem reduces to the Simple Rule of Existence)

In the proof we shall employ, in addition to the Simple Rule of Existence, mainly the following tautologies:

T 1)　　$(p \supset q) \supset (pq \supset r) \supset (p \supset r)$

　　　　(a form of transitivity)

T 2)　　If x is not free in B:

　　　　$B . (\exists x)A = (\exists x)(B.A)$

T 3)　　If x is not free in B:

　　　　$((\exists x) A \supset B) = (x) (A \supset B)$

Proof:

1) $K_1 . K_2 ... K_n . (\exists d_1) L_1 ... (\exists d_M) L_M \supset d_1 = f_1 ... d_M = f_M$

(using T 2) (M-1) times and interchange in 1) ---our hypothesis---we obtain:)

2) $(\exists d_1)(\exists d_2) ... (\exists d_{M-1})(K_1 ... K_n . L_1 ... L_{M-1} . (\exists d_m) L_M) \supset d_1 = f_1 ... d_M = f_M$　　　$\supset$

(by T 3) and interchange in 2)：)

3) $(d_1)\ldots(d_{M-1})$ $(K_1\ldots K_n \cdot L_1\ldots L_{M-1}\cdot(\exists d_M)L_M$ ⊃ $d_1=f_1\ldots d_M=f_M)$

(applying specification to 3) and simplifying the consequent：)

4) $K_1\ldots K_n \cdot L_1\ldots L_{M-1}(\exists d_M)L_M$ ⊃ $d_M=f_M$

(applying the Simple Rule of Existence to 4)；)

5) $K_1\ldots K_n \cdot L_1\ldots L_{M-1}$ ⊃ $(\exists d_M)L_M$

(generalizing 5)：)

6) $(d_1)\ldots(d_{M-1})(K_1\ldots K_n \cdot L_1\ldots L_{M-1}$ ⊃ $(\exists d_M)L_M)$

(by T 3) and interchange in 6)：)

7) $(\exists d_1)\ldots(\exists d_{M-1})$ $(K_1\ldots K_n \cdot L_1\ldots L_{M-1})$ ⊃ $(\exists d_M)L_M$

(by T 2) and interchange in 7)：)

8) $K_1\ldots K_n \cdot (\exists d_1)L_1 \cdot (\exists d_2)L_2 \ldots (\exists d_{M-1})L_{M-1}$ ⊃ $(\exists d_M)L_M$

(The following is a substitution instance of T 1)：)

9) (line 8) ⊃ ((line 1) ⊃ $(K_1\ldots K_n \cdot (\exists d_1)L_1\ldots(\exists d_{M-1})L_{M-1}$ ⊃ $d_1=f_1\ldots d_M=f_M))$

(using modus ponens, and simplifying by dropping $d_M=f_M$：)

10) $K_1\ldots K_n \cdot (\exists d_1)L_1\ldots(\exists d_{M-1})L_{M-1}$ ⊃ $d_1=f_1\ldots d_{M-1}=f_{M-1}$

(From line 10) we get, using the induction hypothesis:)

11) $K_1 \ldots K_n \supset (\exists d_1)L_1 \ldots (\exists d_{M-1})L_{M-1}^-$

(This is a substitution instance of T 1):)

12) (line 11) $\supset$ ((line 8) $\supset$ ($K_1 \ldots K_n \supset (\exists d_M L_M)$))

(modus ponens in 12):)

13) $K_1 \ldots K_M \supset (\exists d_M)L_M^-$

(merging the implications 11) and 13):)

14) $K_1 \ldots K_n \supset (\exists d_1)L_1 \ldots (\exists d_M)L_M$

$$q.e.d..$$

With this proof, the task of formalizing the elementary calculus is complete. We have formally developed the notation of the system. We have stated its axioms, including the Principle of Extensionality; and we have interpreted the Rule of Existence (and derived the stronger form of the Rule as a metatheorem). With this rule, our enumeration of the primitive assumptions of the system is exhausted.

The Relative Frequency Interpretation

The frequency interpretation of the Calculus of Probability is essentially similar to the frequency interpretation presented in chapter I for infinite sequences.

The fundamental difference is that, since we regard all probability sequences as finite, references to the 'limit' are dropped. Hence probability is defined as simply relative frequency. (For example, let us suppose that in a sequence of length 100, 30 of the members of it are B's. Then the probability of a B i this sequence is said to be simply the relative frequency $\frac{30}{100}$.)

Why we choose this interpretation, we discuss below. For the moment we concern ourselves with the question of its admissibility.

In order to show that this interpretation makes the axioms true statements (in fact, tautologies) we employ the notation we introduced in the first chapter, in which we can write:

18) $\quad Fn(A,B) = \dfrac{N(x_1 \ e \ A, y_1 \ e \ B)}{N(x_1 \ e \ A)}$

We shall further simplify our notation by writing the right side of this equation as:

18') $\quad \dfrac{N(A.B)}{N(A)}$

Then our interpretation assumes the following form: if the length of the sequences mentioned in the formula is n, then $(A_1 \underset{c_1}{-} B_1)$ has the same meaning as $Fn(A_1 B_1) = c_1$; or using the P-notation:

19) $\quad P(A_1, B_1) = Fn(A_1, B_1) = N(A_1.B_1)/N(A_1)$

Axiom I (univocality) is immediately seen to be true under this interpretation, since it asserts that the quotient 18') has a unique value just in case that the denominator is not zero. Axiom II, 1.(normalization) assumes the following form in our interpretation:

20) $(A \ni B) \ni N(A.B)/N(A) = 1$

But when $(A \ni B)$, the number of values of i for which A.B holds is the same as the number for which A alone holds. That is, in view of

21) $(A \ni B) \ni (\equiv A.B)$

by interchange we immediately obtain:

22) $(A \ni B) \ni N(A) = N(A.B)$

which is seen to be equivalent to 20).

The second part of this group of axioms (II,2.), states merely that the quotient 18') is non-negative whenever the denominator is not zero. This assertion follows immediately from the non-negative character of the 'N' symbol, under our definition.

The Theorem of Addition may be stated thus (in the P-notation):

23) $P(A, BvC) = P(A,B) \neq P(A,C)$

where the hypothesis is understood that $A.B \ni \overline{C}$. In our notation the interpretation of 23) reads:

24) $\dfrac{N(A.BvC)}{N(A)} = \dfrac{N(A.B)}{N(A)} \neq \dfrac{N(A.C)}{N(A)}$

or equivalently:

25) $N(A.BVC) = N(A.B) \neq N(A.C)$

This follows immediately from the following considerations: when P and Q are incompatible attributes, we have from the definition of the''N' symbol:

26) $N(PVQ) = N(P) \neq N(Q)$

Furthermore:

27) $A.BVC = A.BVA.C$ (distributive law for '.' and 'V')

28) $(A.B \supset \bar{C}) \supset (A.B \supset \overline{A.C})$ (a tautology from the propositional calculus)

From 26) we have:

29) $(A.B \supset \overline{A.C}) \supset N(A.BVA.C) = N(A.B) \neq N(A.C)$

From this 25) follows in view of 27) and 28).

We have thus proved Axiom III. Axiom IV is the following statement (in the P-notation):

30) $P(A,B.C) = P(A,B)P(A.B,C)$

Its interpretation is this:

31) $\dfrac{N(A.B.C)}{N(A)} = \dfrac{N(A.B)}{N(A)} \cdot \dfrac{N(A.B.C.)}{N(A.B)}$

which, we are almost ashamed to point out, is simply the rule for multiplying two fractions!

This completes the derivation of the axioms of the Elementary Calculus of Probability from the relative frequency interpretation.

This completes the derivations of the axioms of the Elementary Calculus of Probability from the relative frequency interpretation.

Turning now to the grounds for adopting this interpretation (besides its admissibility), we recall that we have held that any justifiable interpretation of probability must have this characteristic: that we must be able to show that it is a rational policy to wager on an event. because it has a high 'probability'. In the relative frequency interpretation this can be done. For, we can immediately argue that the relative frequency is simply the percentage of our wagers that we shall win if we keep betting that an event has a certain attribute. Thus, a high 'probability' means a high percentage of successes for the strategy of betting that the event will have. the attribute with that 'probability'. But this is to say that it is a reasonable strategy to posit that the event we are interested in predicting has the attribute. (We remind the reader of the connection between 'positing' and 'winning strategy' that we defined in Chapter I.)

This argument bears a great formal resemblance to the corresponding argument advanced for the frequency interpretation in the case of infinite sequences in Chapter I. Indeed, the only genuine difference in the two arguments is this: where before we had to speak of the success-ratio, or 'long-run frequency of successes', we are able now (in view of the finiteness) to speak more simply of the 'percentage of successes'. But this change only strengthens the argument, inasmuch the reference to the long run is dropped. Thus, the objection which might be raised to the infinite justification (perhaps best expressed by Lord Keynes' 'in the long run we'll all be dead.') analogously to the objection raised against the justification of induction, cannot be made here.

This completes one part of the justification of our choice of an interpretation. But another part remains: to show that induction can determine probabilities, that is, that it can find relative frequencies. This we postpone until we come to take up the general question of induction and its justification in the finite case.

The Theory of Order

In addition to the part of the Calculus of Probability so far discussed, there is a second set of concepts and axioms, constituting the 'Theory of Order', required for the development of the mathematical theory. Whereas we have so far regarded probability sequences as unanalyzed wholes, characterized completely by their probabilities for our purposes, we now investigate their structure.

This description of the structure of probability sequences remains purely formal, however, in this sense: the characterization employs only probabilities, and hence remains within the uniterpreted calculus of probabilities.

More precisely: whereas we have so far described se-

quences only in terms of the 'major probabilities' P(A,B)
we shall now describe them more explicitly by speaking of
the probabilities in subsequences of the original sequence.
This will permit us to describe, for example, some kinds
of randomness in sequences in purely formal terms.

The problem of laying down definitions of such concepts as 'random' within the formal theory, represents
part of the problem of the Theory of Order. In addition
to characterizing different possible types of order in
its definitions, however, it lays down certain axioms valid for all sequences.

These axioms are concerned with phase probabilities.[67]
As an example of what we mean by a 'phase probability',
we may cite the probability that, if a certain member of
a sequence is a B, the next member will also be a B. In
the P-notation this may be symbolized:

32) $P(x_1 \text{ e } A.y_1 \text{ e } B, y_1 \not{/} \overset{.}{1} \text{e } B)$

Where a 'phase' $\neq$ 1 occurs in the subscript. We
shall abbreviate formulae like 32) by writing the phase
as a superscript, thus[68]:

33) $P(A.B, B^1)$

Similarly, the probability of getting a B, followed
by a $\bar{B}$, followed by another B; if we have already encountered a B, is written:

34) $P(x_1 \, e \, A . y_1 \, e \, B . y_1 \neq_1 \, e \, B . y_1 \neq_2 \, e \, \bar{B}, \; y_1 \neq_3 \, e \, B)$

and abbreviated:

35) $P(A.B, B^1 \bar{B}^2 B^3)$.

Now, the purpose of the axioms of the Theory of Order
is to allow us to make certain transformations with phase
superscripts. First of all, we are allowed to drop a
phase superscript attached to a compact reference class (a
reference class that contains all the x_i). That is, if A
is compact:

36) . $P(A^a, B) = P(A, B)$

Secondly, if A is compact, and all the classes are
infinite (that is, if infinitely many members of the corresponding sequence belong to each class); we are allowed
to change all the phase superscripts by the addition or
subtraction of a constant. Thus:

37) $P(A, B^1 \bar{B}^2 \, B^3) = P(A, B^{1 \, \neq \, 3} . \bar{B}^{2 \, \neq 3} . B^{3 \, \neq \, 3})$

        $= P(A, B^4 \bar{B}^5 B^6)$

        or subtracting a constant:

        $= P(A, \bar{B}^{-1} \bar{B} \, B^1)$

If phase superscripts occur before the comma, the
addition must also be made there (except in the case of
'A'. But in view of 36) phase superscripts with 'A' need
not be written at all.) For instance:

38) $\quad P(A.B^5,C^6) = P(A.B^{10},C^{11})$

but not:

$\quad P(A.B^5,C^6) = P(A.B^5,C^{11})$

Our problem, in finitization, is to find an interpretation of these axioms which will legitimize the manipulations they permit (adding and dropping phase superscripts) in the case of finite sequences.

The problem of defining randomness is one that has been investigated with noteworthy results by Richard Von Mises[69]. His work is based on a central characteristic of sequences like those obtained in the study of games of chance; namely, that there is no way to 'beat the game'.

That is, if the probability of an event in the sequence of 'plays' in the game is p, then the probability of the event in any subsequence selected according to a rule depending only on the number of the event and the nature of the preceding events, is also p. For example: The probability of an ace, in the sequence of events obtained by tossing an unbiased die, is 1/6. It will not help us if we bet that an ace will come up only when ten aces have already come up (or only when an ace has not come up for a hundred throws). For the probability of an ace coming up in the subsequence of places selected from the original sequence by the occurrence of aces in the preceding ten places (or by the occurrence of no aces in the preceding hundred places) is still just 1/6.

Thus, the identity of the phase-probabilities, and
the probabilities in these sequences, with the major probability, expresses the fact that no 'game system' will
succeed. It is this---Von Mises terms it the 'Principle
of the Excluded Game System'[70]---that provides Von Mises
with his definition of randomness.

But Von Mises' definition is too strong and leads
in fact to contradictions[71]. Church[72] has formulated this
definition in a logically satisfactory manner, however:
namely, by demanding that no subsequence, selected from
the original sequence by a _recursive_ function depending
only on the place in the sequence and the events in the
preceding places, possess a probability different from
the major probability. Sequences satisfying this condition, we shall call 'logically random'[73] sequences.

While we are interested in studying the properties
of such sequences, where we differ with Von Mises (in
this we follow Reichenbach[74]), is in not restricting
the Theory of Probability to the study of such 'random'
sequences.

Such a restriction appears unnatural, in the first
place, because the class of sequences that satisfy the
axioms of our calculus (under the frequency interpretation, which Von Mises accepts) is very much wider than
just the random sequences. The reason that Von Mises
makes such a restriction is, in part, that he takes as

the archetype of probability sequences, the sequences obtained from games of chance. But there are other sequences, not possessing this character of 'logical randomness', which have great physical significance, and which may also be described in terms of certain probabilities.

As an example of such a sequence, we may consider the following case[75]: Let us suppose that we have a box that contains a gas, and that all of the gas is restricted to one half of the box by a thin partition in the center. Now we open a small hole in the partition. If we consider the position of a given molecule of the gas at successive instants (that is, at time-points a certain short interval apart), we obtain a sequence of positions:

$$x_1, x_2, x_3, \ldots$$

In this sequence we consider each member with respect to the attribute B, defined in this way: B is the property of being in the half of the box that originally contained the gas. Evidently x_1 is a B (since it must have originally been in this half of the box); and we may expect that a long initial section of this sequence will consist of B's. Eventually, however, we shall find a $\bar{B}$ (since the molecule will sooner or later get into the other half of the box), probably followed by a string of $\bar{B}$'s (since the molecule tends to remain in whichever half it may be).

The probability of a B in this sequence is simply

1/2. For in the long run the molecule will be in the original half of the box one half of the time. But the sequence certainly does not satisfy Von Mises' criterion: for getting a B strongly influences the probability of getting another B. That is, if I wait until a B occurs, and then bet that the next member will be a B, I shall win more than half of the time. (In fact, I shall win most of the time.) This type of sequence we may call an 'after-effect' sequence.

Since, under the frequency interpretation as presented in Chapter I, we have a vast range of sequences satisfying the theory (namely, all the sequences that possess a limit of the frequency); including the random sequences and, at the other end of the spectrum, the after-effect sequences, we can be far more lenient than Von Mises in our definition of a probability sequence. In our theory, we shall have, in the first place, theorems asserted for all probability sequences of any type; and secondly, theorems asserted for the sequences of each of the special types.

Our decision for a sweeping definition of probability sequences has the following formal advantage: that we do not assert any theorem upon hypotheses stronger than actually required to insure the validity of the theorem. This consideration is analogous to an argument advanced by Quine[76] for a similar latitude in defining the formulae of

Quantification Theory (and the propositional calculus).
According to Quine[77], we should allow any statement as a
substituent for the sentential letters, and any sentential
schema as a substituent for the functional letters of
Quantification Theory; because the addition of such restrictions as are imposed by the theory of types, while
necessary for set theory, is not required for Quantification Theory (which has no difficulty with consistency).

Quine's argument is that we should introduce restrictions on the statements we consider in logic only
when they are actually required by the level on which
we are working. Similarly, Reichenbach argues that we
should introduce restrictions on the sequences we consider only when they are required for the proof of our
theorems.

In defining random sequences, it is not necessary
to say, with Church-Von Mises, that there is no recursive function that 'beats the game'; it is sufficient that
it be technically impossible to find such a function (that
is, to pick out a subsequence with different properties
from the major one). Sequences with this property, we
shall call 'psychologically random'.

Many of the game-theoretical purposes for which the
concept of 'logical randomness' was developed are satisfied by 'psychological randomness'.

Still more general than the random sequences, are

the normal sequences of Reichenbach[78]and Copeland[79].

Normal sequences, while they may be predictable (un-
like random sequences), share many of the mathematical
properties of random sequences. Thus, most of the theorems developed for random sequences (like the various
forms of the Special,Theorem of Multiplication[80], or
the Bernoulli Theorem) hold for all normal sequences.
Normalcy, indeed, represents the main randomness condition required in the Theory of Probability.

So far, we have spoken only of infinite sequences in
this connection. When we come to finite sequences, we immediately see that there are no logically random sequences
(since every finite sequence may be recursively defined,
because it can be finitely enumerated). There are, however, sequences similar to normal sequences, and we shall
show that it is possible to develop a finite analogue to
the theory of normal sequences.

The Formalization of the Calculus of Probability:
The Theory of Order

The Axioms of the Theory of Order[81] are the two statements we list below. In writing these statements we employ
the abbreviations ($\exists\,\infty$ C) to mean that there are infinitely many members of the corresponding sequence in the
class C. The notation of phase superscripts has already
been explained[82]; and, as usual, we do not write the se-

quence variables or running subscripts.

V. AXIOMS OF THE THEORY OF ORDER

1. $(1)(x_1 \ e \ A).(\exists \ \mathcal{A} \ C) \ \Rightarrow \ (A^a.C \ \underset{p}{\Leftarrow} \ B^s) \ \Rightarrow (A.C \ \underset{p}{\Leftarrow} \ B^s)$

2. $(1)(x_1 e \ A).(\exists \ \mathcal{A} \ C^a...G^s) \ \Rightarrow$
    $(A.C^a...G^s \ \underset{p}{\Leftarrow} B^t) \ \Rightarrow (A.C^{a-r}...G^{s-r} \ \underset{p}{\Leftarrow} B^{t-r})$

In the second axiom, the clause $(\exists \ \mathcal{A} \ C^a...G^s)$ means that there are infinitely many positive integers i such that: $(z_i \ \ne \ _a e \ C...w_{i/s} \ e \ G)$. These axioms have the same meaning as the following formulae in the P-notation (when the appropriate hypotheses are fulfilled):

39) $P(A^a.C,B^s) \ \Rightarrow \ P(A.C,B^s)$
40) $P(A.C^a...G^s,B^t) \ \Rightarrow \ P(A.C^{a-r}...G^{s-r},B^{t-r})$

In the finite interpretation, the clause $(\exists \ \mathcal{A} \ C)$, and the corresponding clause in V. 2., are, of course, always false (because all the sequences are finite). For this reason, the axioms of the theory are immediately true (trivially, because of the always-false antecedent), but empty. It would thus appear that we have no difficulty in showing that the finite frequency interpretation satisfies the Theory of Order.

But such a demonstration is too dearly bought. It is meaningless to show that formulae 39) and 40) may be applied whenever the hypotheses are true, if the hypothe-

ses are never true. What we really desire is to show that
39) and 40) may be applied to the widest possible class
of probability sequences.

In the development of the finite theory, it accordingly becomes necessary to show that we can assert these equations without such strong preconditions. In other words,
it becomes necessary to replace the axioms V. by the
stronger axioms obtained by dropping the clauses which assert the infinity of certain sequences. This amounts to
asserting 39) and 40) whenever the class A is compact.
(This strengthens the theory, since the original axioms
are subalterns of the new ones).

Before attempting to show that this can be done, it
is necessary to examine more closely the meaning of the
clause (i) (x_i e A), which we have so far been using to
express compactness. Actually, this says more than merely 'all the x's are A's'. It asserts that for every positive integer (since we regard the range of the variable i
as the positive integers) there is a corresponding member
of the x-sequence belonging to the class A. If we make
the natural assumption that all the x_i are distinct, this
is true only if the x-sequence is infinite.

The important fact behind this is that the 'sequence-
variables' are not variables at all (this is reflected in
the fact that quantification of sequence or element variables is not permissible), but descriptions:'x_1' is short

for 'the i-th member of the x-sequence'.

The convention we have adopted in the case of the infinite theory is to regard the statement 'x_i e A' as false whenever the number of elements in the sequence is some positive integer n, and i n. This convention is a semantical one: it does not constitute an axiom, or one of the rules we employ in making deductions from the axioms, but it lays down a meaning for some of the expressions of the system. If we are to use the simple expression (i) (x_i e A) to express the compactness of A, we shall have to modify this convention, when we come to speak of finite probability sequences.

The convention we propose is the following: In speaking of a sequence of length 10, for example, we shall regard 'x_1', 'x_2',...,x_{10}' as designating the first, second,...,tenth member of the sequence (as before), and 'x_{11}' as designating the first member, 'x_{12}' the second member,...,'x_{20}' the tenth member, 'x_{21}' the first member again,...etc. This procedure we call, for obvious reasons, 'circular counting'.

The effect of this convention is to assign a member of the x-sequence, x_i, to every positive integral value of i. As a result 'Σ(i) (x_i e A)' comes to be true whenever all of the x-sequence belongs to A.

In order to assign a designation to x_i for negative values of i, which we desire to do in order to carry out the proofs below, we extend our circular counting to the reverse direction. Taking a sequence of length 10, as before, this means that x_0 designates the last member of the sequence (x_{10}), x_{-1} designates its predecessor x_9, x_{-2} designates x_8 and so on.

This convention, like the one adopted in the infinite case, is not a part of the formal system: its effect is to reinterpret the formulae of the system so that certain statements come out true when we speak of finite sequences. We can, however, state a corresponding formal rule. In order to do this we employ the 'congruence' notation of mathematics, writing

$$a \approx b \pmod{n}$$

for: 'n divides a - b' or 'a and b leave the same remainder on division by n'. Our rule, which we shall call 'Convention 0', may be stated thus (where b represents a sequence variable):

CONVENTION 0

If the b-sequence is of length n, then:

$$b_j \approx b_k \text{ if and only if } j \approx k \pmod{n}$$

With this convention in force, we now proceed to show that 39) and 40) hold for finite sequences under the relative frequency interpretation.

To prove 39) we begin with the identity:

41) $P(x_{1/a} \text{ e } A, \, \pi \text{ e } C, \, y_{1/b} \text{ e } B) = P(x_{1/a} \text{ e } A, \, \xi_1 \text{ e } C, \, y_{1/b} \text{ e } B)$

Since all the x_i belong to A, for every positive i and for every a, positive or negative, we have both $x_i e$ A and x_i/a e A. Hence:

42) x_i e A = x_i/a e A

Using 42) and the Principle of Extensionality[83] to put 'x_i e A' for 'x_i/a e A' in the left side of 41) we obtain 39).

To prove 40), we write:

43) $P(A.C^a...C^s, B^t) = \dfrac{N(A.C^a...G^s.B^t)}{N(A.C^a...G^s)}$

For every value of i such that:

x_i e A.z_i/a e C...w_i/s e G.y_i/t e B

we can find a corresponding j such that:

x_j e A.z_j/a_{-r} e C...w_j/s_{-r} e G.y_j/t_{-r} e B

namely, $j = n -/r/i$, where n is the length of the sequences. For, by Convention O, this follows in view of

$j-r = i \pmod n$

while the subscript of 'x' does not matter in view of 39).

We have then:

44) $\displaystyle\mathop{N}_{i=1}^{n} (A.C^a...G^s.B^t) = \mathop{N}_{j=n/r/1}^{2n/r} (A.C^{a-r}...G^{s-r}.B^{t-r})$

(where we suppose j to occur instead of i in the right side of 44)).

In view of Convention O, we may imagine the values of j to be replaced by their residues modulo n. In that case, instead of taking on all values from $n/r/1$ to $2n/r$, j will simply take on all the values of the residue class, or $0,1,\ldots n-1$. Since $x_0 = x_n$, we may also rewrite the limits of counting in the right side of 44) as $j=1$ to n. Doing this, and putting i for j, we finally obtain:

45) $$\prod_{i=1}^{n} (A.C^a \ldots G^e.B^t) = \prod_{i=1}^{n} (A.C^{a-r} \ldots G^{e-r}.B^{t-r})$$

In the same way we get:

46) $$\prod_{i=1}^{n} (A.C^a \ldots G^e) = \prod (A.C^{a-r} \ldots G^{s-r})$$

But we can write:

47) $$P(A.C^{a-r} \ldots G^{s-r}, B^{t-r}) = \frac{N(A.C^{a-r} \ldots G^{e-r}.B^{t-r})}{N(A.C^{a-r} \ldots G^{s-r})}$$

and from 47) and 43) we obtain 40), in view of 45) and 46).

In our proof of 40) we had to assume, in addition to the compactness of A, that the B-,C-,....,G-sequences were all the same length. The statement that two sequences, for example the B-sequence and the C-sequence, are the same length may be symbolized thus: $E(B,C)$, which we shall construe as an abbreviation for:

48)   (n)   (i) (j) $(y_i = y_j . =. \ i = j \pmod n))$ $\supset$
            (i) (j) $(z_i = z_j . =. \ i = j \pmod n))$

Similarly, we can define $E(B,C\ldots G)$ which we shall use to express the equality of these sequences with respect to length. With this notation, the axioms of the

Theory of Order may be written:

V'. AXIOMS OF THE FINITE THEORY OF ORDER

1.   (i) $(x_i \in A) \supset (A^a.C \underset{p}{\sim} B) \equiv (A.C \underset{p}{\sim} B)$

2.   (i) $(x_i \in A) . E(B.C...G) \supset$

       $(A.C^a...G^s \underset{p}{\sim} B^t) \equiv (A.C^{a-r}...G^{s-r} \underset{p}{\sim} B^{t-r})$

Types of Order

The theory of the types of order in finite sequences closely parallels the infinite theory, once a technique for modifying the definitions of the infinite theory, to make them fit the finite case, has been worked out. The most important problem for the development of the calculus is the definition of 'normal sequence'. We will illustrate the finite Theory of Order by showing how it is possible to find a finite analogue for normal sequences.

As a preliminary, we must define 'freedom from after-effect'. In a sequence that is free from after-effect, the occurrence of a B does not effect the probability that the next element will be a B. Thus:

$$P(A,B) = P(A.B,B^1)$$

As this equation illustrates, freedom from after-effect is characterized by the fact that the phase probabilities are equal to the major probability.

For the expression of the general case, we introduce, instead of the disjunction $B\bar{V}3$, the many-term disjunction $B_1V...VB_t$, which we suppose to be complete and exclusive.

Then the definition of a <u>sequence free from after-effect</u>[84] is:

49) $\quad P(A.B_{i_1}^{1} \ldots B_{i_{v-1}}^{v-1}, B_{iv}^{v}) = P(A, B_{iv})$

where i_r means: the subscript belonging to the term with the phase r.

It is easy to show that no finite sequences (except the degenerate sequences in which all of the probabilities are zero or one, that is, the sequences $B_1 B_1 B_1 \ldots B_1$) satisfy 49). To show this, we consider a sequence of length n (where n is also the length of the sequence). From our circular counting, it also follows that the 'next' element in the sequence must be the first element in the section. That is:

50) $\quad P(A.B_{i_1}^{1} \ldots B_{i_n}^{n}, B_{i_1}^{n+1}) = 1$

If the section $B_{i_1} \ldots B_{i_n}$ actually occurs somewhere in the sequence, then we can also show that the probability from the section to B_j, $j \neq i_1$, is 0. From this our assertion follows.

If we had not introduced our convention of circular counting, we could still prove that no finite sequences satisfy 49). For then we would have, in a sequence of length n, $P(A.B_{i_1}^{1}, B_{i_2}^{2} \ldots B_{i_{n-1}}^{n-1}, B_{i_n}^{n}) = 0$ (whatever the value of i). This is true because $x_{i \neq n} \in B_1$ is false, whatever value we give i, (because there is no element $x_{i \neq n}$).

But then, if the sequence were free from after-effect, we would conclude with the aid of 49) that

$$P(A, B_i) = 0 \quad \text{for every } i$$

which is impossible, because the sum of the probabilities of the B_i has to equal one.

In general, it is evident that the occurrence of a certain combination in a sequence can leave unaltered the probabilities of the succeeding elements only if the combination is much shorter in length than the sequence as a whole. This is so because a combination of great length can occur in a finite sequence only a limited number of times. Hence, if we know that a certain place follows such a combination, we can narrow down the number of attributes which may occur in that place, and improve our prediction.

Furthermore, even when v is small, we can in general expect the equation 49) to hold only approximately. This much we can find, however: we can find finite sequences that satisfy 49) to some high degree of approximation d, whenever $v \angle$ some number N_1. Such a sequence we shall call free from after-effect to the degree N_1.

The probability of a combination in a sequence which is free from after-effect to the degree N_1 can be calculated by the Special Theorem of Multiplication[85], provided $N_1 d$, the degree of approximation multiplied by N_1, is small enough and the combination does not exceed N_1 in length. For:

51)
$$P(A,B_{i_1}^1 \ldots B_{i_v}^v) = P(A,B_{i_1}^1 \ldots B_{i_{v-1}}^{v-1}) \cdot P(A,B_{i_1}^1 \ldots B_{i_{v-1}}^{v-1}, B_{i_v}^v)$$

$$= P(A,B_{i_1}^1 \ldots B_{i_{v-1}}^{v-1})(P(A,B_{i_v}) \not< d) \qquad \text{(employing 49))}$$

$$= P(A,B_{i_1}^1 \ldots B_{i_{v-1}}^{v-1}) \cdot P(A,B_{i_v}) \not< d \qquad \text{(making use of the}$$

fact that probabilities are $\not< 1$.[86])

After v-1 more steps (repeatedly applying the same procedure to the first term on the right side) we find:

52)
$$P(A,B_{i_1}^1 \ldots B_{i_v}^v) = P(A,B_{i_1}) \ldots P(A,B_{i_v}) \not< vd$$

This fact is of great practical importance. For the sequences with which we normally deal are of great length so that even very long combinations are still extremely short in comparison to the sequences. It is this that allows us to apply the Special Theorem of Multiplication to the calculation of probabilities in so many cases. Indeed, the degree of freedom from after-effect of these sequences is so high that it may be regarded, for all practical purposes, as infinite.

In defining 'normal sequence', we demand freedom from after-effect and one thing more: that the probabilities (including phase probabilities) in the regular subsequences be equal to the major probabilities. By the regular subsequences, we mean the subsequences of the form: the kth element and every l'th element thereafter.

(E.g., $x_5, x_8, x_{11}, x_{14}, \ldots$). We will denote such a regular
subsequence by S_{1k}, where k is the number of the first e-
lement belonging to the subsequence and 1 is the distance
between elements. Thus S_{1k} consists of all the x_i for
which i=k (mod 1). That a member, x_1, of the sequence belongs to the subsequence we indicate by writing $x_i \in S_{1k}$.
The probability that a member of the subsequence is a B_1
is written $P(A.S_{1k}, B_1)$; phase superscripts may also be
employed with the symbol in the usual manner. Using this
symbol, we write the relation defining normalcy thus:

53) $P(A.S_{1k}, B_1) = P(A, B_1)$

$P(A.S_{1k}.B_{i1}^1 \ldots B_{i_{n-1}}^{n-1}, B_{i_n}^n) = P(A.B_{i1}^1 \ldots B_{i_{n-1}}^{n-1}, B_{i_n}^n) = P(A, B_{i_n})$

$1 \leq a \leq n$ $1 \leq k \leq 1$

The problem of normalcy for finite sequences is analogous to the problem of freedom from afetr-effect. Evidently no finite sequence satisfies 53) (excepting degenerate cases), for choosing k=1, '1'=n (the length of the sequence), we have:

54) $P(A, S_{n,1}, B_1) = 1$ or 0 as B_1 is the attribute of the
of the first member of the sequence or not.

This is true because $S_{n,1}$ is the subsequence consisting of the single element x_1. If x_1 is a B_k, then
$(1)(x_1 \in S_{n,1} \supset x_1 \in B_k)$, and from this, with the aid of
the axiom of Normalization, we obtain 54).

There are, however, sequences which satisfy the conditions 53) to a high degree of approximation whenever $n,1 \angle N_1$. Such sequences we shall call _normal to the degree N_1_. For normal sequences the Special Theorem of Multiplication can be used in determining the probability of the occurrence of a combination (not merely anywhere in the sequence, but in a system of non-overlapping sections[88]) provided the degree of approximation is sufficiently good.

The proof is given by Reichenbach[89]. The modifications required for the finite case are similar to those made in 51)-52).

The Uniqueness of the Interpretation

In a preceding section[90] we have argued on material grounds for the adoption of the relative frequency as the interpretation of probability in finite sequences. These grounds were that the adoption of this interpretation makes our employment of probability concepts consistent with our actions; that is, that it makes it possible to explain why a knowledge of the probabilities of future events is a guide to action.

With any formal system, however, we face an additional problem. Beyond providing the system with a natural and acceptable interpretation, we wish to discover how wide the class of admissible interpretations actually is. In the finite theory of probability, we shall show that

we have an extreme case: there is only one admissible
interpretation; namely, the one we have given.

To show that the interpretation is fixed by the axioms, we shall show that <u>within the uninterpreted system</u>
we can prove that the probability in a finite sequence is
equal to the relative frequency. In view of the very general nature of the axiom set, and the multiplicity of interpretations in the infinite case[91], <u>this a very</u> surprising feature of the finite theory.

We suppose, accordingly, that we have a finite sequence
$x_1, \ldots, x_n$. In this sequence we suppose that there are
m B's. The class of the x_i we term A (so defined, A is
evidently compact), and we seek to prove that the probability of B is equal to the relative frequency m/n.

We shall use the symbols $C_1, \ldots, C_i, \ldots$ to denote the
classes whose sole members are $x_1, \ldots, x_i, \ldots$ respectively
(in other words, C_i is the unit class of x_i).[92]

Let the elements $x_{1_1}, \ldots, x_{1_m}$ be the m B's. Then
an element is a B only if it belongs to one of $C_{1_1}, \ldots, C_{1_m}$.
In symbols:

55) $(B \cdot C_{1_1} V C_{1_2} V \ldots V C_{1_m})$

Hence, if all the probabilities exist:

56) $P(A,B) = P(A,C_{1_1}) \neq \ldots \neq P(A,C_{1_m})$

(since the C_i are disjoint by definition). Furthermore,

all of the x's belong to some C_i (in fact, each x belongs to the corresponding C-class):

57) $(A=C_1V....C_n)$

By the Axiom of Normalization:

58) $P(A,C_1V...VC_n) = 1$

and if all the probabilities exist:

59) $1=P(A,C_1V...VC_n) = P(A,C_1) /..../P(A,C_n)$

If we can prove that all of the $P(A,C_i)$ are equal, it is an immediate consequence of 59) that they all equal 1/n (because there are n of them, and they total 1); and of 55) that $P(A,B)=m/n$ (Because it is the sum of m of them, and they each equal 1/n).

By the Theory of Order:

60) $P(A,C_n) = P(A,C_n^I)$.

But $x_{i/1}eC_n = x_ieC_{n-1}$ (by the definition of the C_i). In our abbreviated notation:

61) $(C_n^I=C_{n-1})$

By the Principle of Extensionality:

62) $P(A,C_{n-1})=P(A,C_n^I)=P(A,C_n)$

Thus we have proved the equality of $P(A,C_n)$ and $P(A,C_{n-1})$.

Similarly, we can prove the equality of $P(A,C_{n-1})$ and $P(A,C_{n-2})$, and in n-1 such steps we obtain:

63) $P(A,C_n)=P(A,C_{n-1})=\ldots=\ldots=P(A,C_1)$

We have shown, thus, that the probabilities $P(A,C_i)$ have, if they exist, the value 1/n. Therefore they exist (by the Rule of Existence, since they are uniquely determined from 0 probabilities). Similarly, we have that $P(A,B)$ exists, and

64) $P(A,B)=m/n$

Thus: 1) every finite sequence is a probability sequence. 2) The probability controlling it is simply the relative frequency.

CHAPTER III

INDUCTION

Introduction

The problem of constructing a satisfactory theory of induction in purely finite terms divides into two major parts. On the one hand, we wish to show that the mathematical requirements of such a theory can be met. For this reason, we have had to show that the theory of probability can be given an acceptable interpretation for the finite case. For the same reason, we shall have to show that certain theorems employed in the evaluation of inductive inferences can be derived within the calculus we have developed for the finite case, using the definitions we have proposed for this case. We shall, in fact, have to do more than merely derive these theorems for this case. If this were all we had to do, this part of our task would amount to no more than an extension of the work we have done in the last chapter; it would remain entirely within the formal theory. But we must show how the theorems in question can actually be applied to finite sequences.

The theorems in question are concerned with the appraisal of inductive inferences. They concern, therefore, 'advanced knowledge', i.e., that state of knowledge in which we have not merely a knowledge of probabilities

obtained by the use of the Rule of Induction, but a knowledge of the character of the sequences themselves (e.g.,
a knowledge that they are normal sequences), and a knowledge of the probability that the relative frequency in
these sequences will assume any given value.

This last knowledge is a knowledge of the antecedent
probability[93] (the probability before any examination of
the particular sequence has been made) that the probability in that sequence lies in a given interval. As a
'probability of a probability', it is a probability of a
higher level. In particular, it is a second level antecedent probability. It is also, like the first level probabilities, a probability in a sequence; but in a sequence
of sequences. If we ask for the probability that the
second level probability lies in any given interval, we
are asking for a third level probability, and so on. Thus,
advanced knowledge is characterized by the fact that it is
knowledge of probabilities of a higher level; the peculiar
advantage to being in advanced knowledge is that these
probabilities can be used, provided we have certain
information about our sequences, to appraise our inductive
inferences; to tell us the probability that a particular
application of the Rule of Induction gives us the correct
result.

The theory of advanced knowledge is, thus, partly
formal (insofar as it is concerned with the derivation of

the theorems that enable us to appraise our inductive inferences when we have the requisite knowledge), and partly interpreted (insofar as it is concerned with explaining how the Rule of Induction can lead from primitive to advanced knowledge[94]). Both parts of the theory are mathematical in character; thus the difference here is not between mathematics and philosophy, but rather between pure and applied mathematics.

The other major part of our task is philosophical in character. The nature of this philosophical problem we have already explained: to justify our use of the inductive rule.

In this chapter we shall complete the first part of this task. This involves: 1) proving the Bernoulli theorem for finite sequences, 2) showing how the inductive inference may be appraised when the necessary probabilities are known, 3) showing how the inductive rule can lead from primitive knowledge to advanced knowledge.

By the conclusion of this chapter, then, we hope to have shown how the entire structure of empirical knowledge is built up, using only the Rule of Induction as its basic instrument. What this amounts to is showing that, within the compass of the finite theory, we can give a complete and satisfactory account of statistical inference. We cannot, of course, show that the use of the methods we develop will lead to success; what we can show within the

purely mathematical sphere is <u>how</u> they lead to success
when certain conditions exist. Our justification for
using these methods, or (what amounts to the same thing)
hoping these conditions exist, we shall present in the
next chapter.

The Bernoulli Theorem

Let us suppose that we have a finite sequence which is
normal to a very high degree N^{95}, so that the special
theorem of multiplication can be used for the calculation
of the probability of combinations. We suppose, further,
that the probability controlling the sequence is p.

Our usual method of estimating this probability is by
the use of induction. That is, we find a sample, consisting of some n members of the sequence, and calculate the
relative frequency Fn(A,B). This frequency enables us to
make a posit as to the value of the probability p.

In the case of normal sequences, it is easy to justify this procedure; for we know that, in a normal
sequence, the relative frequency in a sufficiently large
sample (which is still small in comparison with N^{96}) is
approximately equal to the probability, in general. This
we shall prove later. That is, we shall show that we can
calculate the probability that P(A,B) lies in the interval
Fn(A,B)$\neq$e, and we shall show that this probability is
approximately equal to one for large n.

Our question at the moment concerns the reverse of this probability, i.e., the probability that $Fn(A,B)$ lies in the interval p/e. More generally, how often can we expect that $Fn(A,B)$ will lie in a given interval?[97]

First of all, let us inquire as to the probability of obtaining $Fn(A,B)=m/n$. This is equivalent to asking: how likely are we to find m B's among n successive elements?

From the theory of permutations we know that this can happen in $n!/m! \cdot (n-m)!$ ways. ($n!/m!(n-m)!$ is also called the __binomial__ __coefficient__ $\binom{n}{m}$.) The probability of obtaining a particular combination of n elements, m of which are B's, is, by the special theorem of multiplication, equal to the probability of getting a B, or p, raised to the nth power, multiplied by the probability of $\overline{B}$, or (1-p), raised to the (n-m)th power, i.e., $p^m(1-p)^{n-m}$.

Let us denote the occurence of a combination of n elements, m of which are B's, by F_m^n . Thus, we have:

$$F_m^n =_{df} (B^1 B^2 \ldots B^m . \overline{B}^{m/1} \ldots \overline{B}^n) \; v \ldots v \; (\overline{B}^1 \ldots \overline{B}^{n-m} \ldots B^n)$$

in which the terms of the disjunction consist of all the combinations that contain m letters B and n-m letters $\overline{B}$ in any arrangement. Since all the $\binom{n}{m}$ terms of the disjunction have the same probability, we derive:

65) $P(A, F_m^n) = \binom{n}{m} \; p^m(1-p)^{n-m} = w_{nm}$

This expression is Newton's formula, which we have

derived for finite sequences (with approximate equality
replacing strict equality[98]).

Let $b_n(f_1, f_2)$ denote the probability of finding a
value for $Fn(A, B)$ between $f_1 = m_1/n$ and $f_2 = m_2/n$. By 65)
this equals:

66) $$\sum_{m=m_1}^{m=m_2} \binom{n}{m} p^m (1-p)^{n-m}$$

or

67) $$\sum_{m=m_1}^{m=m_2} w_{nm}$$

If we represent $b_n(f_1, f_2)$ by means of a histogram with
$n \neq 1$ divisions (since there are $n \neq 1$ possible values for f,
corresponding to $m = 0, 1, \ldots, n$) the ordinate $w_n(f)$ coordinated
to any point f will be:

68) $$w_n(f) = (n \neq 1) w_{nm} \qquad (f = m/n)$$

As Reichenbach remarks, for large n this can hardly
be distinguished from a smooth curve. Extending the
formula to non-integer values of m and n[98], we can replace
the summation in 67) by an integral, obtaining (compare
The Theory of Probability, p.270):

69) $$b_n(f_1, f_2) = \int_{f_1}^{f_2} w_n(f) df$$

As Reichenbach remarks[100], "This equation, in which the
transition to the continuous distribution is carried through
is strictly valid only for $n \to \infty$, but it represents a very

good approximation for finite n, so that for large n it cannot be distinguished practically from the precise value. We shall follow Reichenbach in calling the functions $w_n(f)$ and $b_n(f_1,f_2)$ a Bernoulli density and a Bernoulli probability respectively.

What we have now proved is that the Bernoulli functions $w_n(f)$ and $b_n(f_1,f_2)$ may be applied to finite sequences exactly as to infinite sequences. The restriction is required (since the special theorem of multiplication is used in the proof) that $n/\!\!\!\!\!\diagup N$. This is not a serious restriction, however, since for practical purposes we never consider sequences as normal unless N is so large that our samples never remotely approach N in magnitude.

The only difference between this proof and the proof in the infinite case is that the formula 65) which holds strictly for infinite normal sequences holds only approximately for their finite counterparts. Since the transition to the continuous distribution we have just carried out replaces the strict equality of 67) by an approximate equality even in the infinite case, this need not trouble us much.

For the infinite case, we have the following property of $b_n(f_1,f_2)$:

70) $\lim\limits_{n \to \infty} b_n(f_1,f_2) = 1$ if $f_1 \diagup\!\!\!\! p \diagup\!\!\!\! f_2$

$\lim\limits_{n \to \infty} b_n(f_1,f_2) = 0$ if $f_1 > p$ or $f_2 < p$

If we introduce the abbreviation:

71) $b_{ne} =_{df} b_n(p-e, p\neq e)$,

the first relation can be written:

72) $\lim_{n \to \infty} b_{ne} = 1$

(evidently, the second of the above relations is derive-
able from the first).

This relation, which we shall refer to simply as
Bernoulli's Theorem[101], has an analogue for finite
sequences. This we shall now develop.

Let us denote B by B_1 and $\bar{B}$ by B_2, and let us coor-
dinate to any element of our sequence the amount $u_1 = 1$ or
$u_2 = 0$ according as it is a B_1 or B_2. Furthermore, let us define the amount of a combination of consecutive elements
by the addition of the seperate amounts. Thus, a combination of n elements that contains m_j elements B gives the
value

73) $u(B_{k_1}^1 \ldots B_{k_n}^n) = u_{k_1} \neq \ldots \neq u_{k_n} = m_j$

(where u_{k_i} denotes the amount coordinated to B_{k_i}, and this
in turn denotes the attribute---B_1 or B_2---possessed by the
ith member of the combination.)

We have as the mean of these amounts[102]:

74) $M(u_{k_1} \neq \ldots u_{k_n})_{k_1 \ldots k_n} = M(u_{k_1})_{k_1} \neq \ldots \neq M(u_{k_n})_{k_n} = n \cdot M(u)$

(This is simply the familiar additive law of the mean. The running subscripts k_i indicate that the mean of u_{k_i} is being calculated for each value of k_i---hence, for all u^{103}.)

But:

75) $\quad M(u) = \sum\limits_{k=1}^{2} P(A,B_k) \cdot u_k = p$

and thus:

76) $\quad M(u_{k_1} \not{+} \ldots \not{+} u_{k_n})_{k_1 \ldots k_n} = M(m_j)_j = M(m) = np$

(where m designates the possible number if B's among the n elements, or $0,1,\ldots,n$) hence:

77) $\quad M(m_j/n)_j = p$

i.e., the average relative frequency is the probability p. (This calculation, which is identical with that for the infinite case, is strictly valid since the assumption of normalcy has not yet been required.)

If we assume the sequence is normal, we obtain in the usual way[104] the additive law for the dispersion (provided $n \angle N$, again). Then we have:

78) $\quad \Delta^2(u_{k_1} \not{+} \ldots u_{k_n})_{k_1 \ldots k_n} = \Delta^2(u_{k_1})_{k_1} \not{+} \ldots \not{+} \Delta^2(u_{k_n})_{k_n}$
$\quad = n\Delta^2(u)$

But:

79) $\quad \Delta^2(u) = \sum\limits_{i=1}^{2} P(A,B_i) \cdot d^2 u_i$

(where $du_1 = u_1 - M(u) = 1-p$; $du_2 = u_2 - M(u) = -p$)

hence,

80) $\Delta^t(u) = p(1-p)^2 \neq (1-p)p^2 = p(1-p)$

and

81) $\Delta^t(u_{k_1} \neq \ldots u_{k_n})_{k_1 \ldots k_n} = \Delta^t(m) = np(1-p)$

82) $\Delta^t(m/n) = \Delta^t(m_j/n)_j = \Delta^t(f^n) = \dfrac{p(1-p)}{n}$

Thus we have shown not merely that the mean of the relative frequencies f^n is the probability (for all sequences), but that for normal sequences the dispersion becomes extremely small as n approaches N.

If we make use of Tchebycheff's inequality[105], we obtain:

83) $b_{ne} \geqslant 1 - \dfrac{p(1-p)}{ne}$

If we assume that $N \cdot e$ is large (or that e is large in relation to $1/N$, which, in view of the magnitude of N, is possible even for quite small values of e) so that we have approximately

$p(1-p)/Ne = 0$

we have:

84): $b_{ne} = 1$ when $n = N$ (approximately)

Thus, in correspondence with the Bernoulli Theorem for infinite sequences, we have the theorem that $b_{n\epsilon}$ approaches and becomes approximately equal to one as n approaches N. (It becomes strictly equal to one, of course, when n equals the length of the sequence.) The relation, in symbols:

85) $\lim\limits_{n \to N} b_{n\epsilon} = 1$

is the finite analogue of 72).

The Inductive Inference

In the preceding section we asked the question: what is the probability that the probability p controlling a s sequence lies in the interval $f_n \angle d$?

We did not answer this question, but we did obtain an answer to the question: what is the probability that f_n lies in the interval $p \angle d$? This probability may be interpreted in several ways. We may take it to refer to the probability that the frequency of B's in any section of n consecutive elements of the sequence equals $p \angle d$[106]; or to the probability that the frequency of B's in a sample consisting of n consecutive elements of the sequence beginning with the $(Kn \angle J)$th element (for every K and some fixed J) lies in the interval (this means dividing the sequence into non-overlapping samples of length n[107]); or to the probability that the relative frequency Fn(A,B) in any of a sequence of sequences, [108]all of which are normal[109] and controlled by the same probability p, lies in the interval.

The question we are now treating is connected with the third of these interpretations. It concerns, then, a set of sequences:

$$s_1, s_2, \ldots, s_i, \ldots$$

We wish to know the probability that the probability p_i controlling any s_i lies in the interval $f_{n_i} \underset{}{\angle} d$ (where f_{n_i} denotes $Fn(s_i, B)$).

This question concerns the inductive inference. For our rule of induction is to posit that $p_i = f_{n_i} \underset{}{\angle} d$. If we can solve this question, then we can determine the probability that the Rule of Induction will give us the correct result from a sample of size n. In other words, we shall be able to <u>appraise</u> the inductive inference[110] (when we possess the requisite information).

Let us, first of all, divide the possible values of p_i into intervals of length $2d=dp$. Let us denote the class of sequences s_i with probabilities in an interval of length dp whose center is p (i.e., with $p_i = p \underset{}{\angle} d$) by $A_{p,d}$. We have, accordingly[111], that

$$P(A.A_{p,d}, B) = p.$$

Let us put:

86) $\quad P(A, A_{p,d}) = q(p)dp$

Thus $q(p)dp$ is the probability that a sequence is controlled by a probability $p_i = p \angle d$, or, in other words, the antecedent probability[112] of p.

If we make the assumption that the subsequence $A_{p,d}$ is a normal lattice[113], we have, by Newton's formula proved in the last section:

87) $\quad P(A \cdot A_{p,d}, F_m^n) = w_{nm} = 1/n \angle 1 \cdot w_n(p;f)$

By Baye's Rule, we obtain the desired probability as

88) $\quad P(A \cdot F_m^n) A_{p,d}) = v_n(f;p)dp$

$$= \frac{P(A \cdot A_{p,d}, F_m^n) q(p)dp}{\sum\limits_{p} P(A \cdot A_{p,d}, F_m^n) q(p)dp}$$

$$= \frac{w_n(p;f) \, q(p)dp}{\sum\limits_{p} w_n(p;f) \, q(p)dp}$$

where the summation is over all of the intervals $p \angle d$ (i.e., p takes on each of the midpoint values).

If dp is sufficiently small, $q(p)$ may be regarded as a continuous function, and 88) becomes:

89) $\quad v_n(f;p)dp = \dfrac{w_n(p;f)q(p)dp}{\int_0^1 w_n(p;f)q(p)dp}$

If we introduce the simplifying assumption that $q(p)dp$ is constant, that is, that the antecedent probabilities are all equal, $q(p)$ drops out in 89), and since

$$\int_0^1 w_n(p;f)dp = 1^{114}, \quad \text{we find:}$$

90) $\quad v_n(f;p) = wn(p;f)$

Thus the Bernoulli function $w_n(p;f)$ has the double meaning, for this case, of the probability that the relative frequency lies in the interval $p_i \nleq d$ (where p_i is the probability controlling the sequence), and also the probability that the probability p_i controlling the sequence lies in the interval $f_{ni} \nleq d$.

Provided we know the antecedent probabilities, that is to say, the function $q(p)$, we can determine, by the use of 89), not merely the probability that $P(s_i,B) = r_{ni}$ (in other words, the probability that the inductive inference is good); but the probability $v_n(f;p)dp$ that $P(s_i,B)$ lies in the interval $p \nleq d$ for _any_ p. We can, accordingly, find that p which maximizes this probability. That is, we can determine the _most probabile_ _value_ of $P(s_i,B)$ on the basis of f_{ni}.

This result constitutes the foundation of modern statistical inference. For the procedure of any statistical inference is to calculate a relative frequency, and to determine, from that frequency, the most probable value of the probability in the total distribution. This can be done 'on the basis of 89) whenever $q(p)$ is known and it is possible to find the maximum value of $v_n(F;p)$.

The greatest limitation of statistical inference is
clearly visible in 89); namely, the necessity for knowledge of the 'apriori distribution' q(p). If we can demonstrate that our rule of induction enables us to find q(p),
and to find when a sequence is normal, then we shall have
shown that all of statistical inference can be developed
within our theory.

We have, furthermore, on the basis of the theorem
proved in the last section[115], the convergence relation:

91) $\lim_{n \to N} v_n(f;p_1,p_2) = 1$ for $p_1 \underline{/} f \underline{/} p_2$

" " " " " " " " " $= 0$ for $f \underline{/} p_1$ or $p_2 \underline{/} f$

This relation holds even without the assumption of
the equality of the antecedent probabilities[116]. We need
only suppose that the probabilities p and q(p) exist, that
q(p) does not vanish for p=f, that the sequences form a
Bernoulli lattice[117], and that p_1-p_2 is large in absolute
value in comparison with $\frac{1}{N}$ [118].

The convergence relation 91) tells us that the values
given by the Rule of Induction represent the probability
controlling the sequence (within a certain interval of
exactness) with a probability v_n; and v_n goes to one as
n approaches N. But as Reichenbach remarks[119]:

"Even this formula...cannot be regarded as supplying
a general justification of the inductive inference, for it
is based on special presuppositions. Its significance lies
rather in the function that it performs in the further
extension of the theory of induction, once the general

justification...has been given.*

The Method of Correction

The problem which remains for us is that of the transition from primitiive to advanced knowledge. We suppose that we have begun by evaluating a great number of probabilities by the use of our rule of induction. These probabilities represent 'anticipative posits'. We now ask how we can transform them into 'appraised posits'.

What we seek to learn is the probability that our posits of the first level are correct. If we find, when we have solved this problem, that some of our posits do not have the highest attainable probability of correctness, then we shall change these posits. Thus, the transition to the higher level is also a process by which our method is used to correct itself.

The logical structure of this 'method of correction' is this: a great number of anticipative posits are made on a given level. The results obtained are used to determine probabilities of the next higher level which enable us to 'rate', or evaluate, our posits the level below. By the repetition of this process, all of our posits on any number of levels can be transformed into appraised posits; the posits of the highest level, however, are always anticipative posits.

This transition may also be described as the

transition from induction by enumeration to more sophisti-
cated statistical induction.

As we remarked in the preceding section, such statistical inference demands a knowledge of the 'apriori distribution', that is, of the antecedent probabilities.
Without a knowledge of these, our only instrument for determining probabilities is the Rule of Induction. Once we
have obtained the antecedent probabilities (and the other
information we require, e.g., the normalcy of our sequences[120]) by the use of this rule, we can employ statistical
methods.

To show, accordingly, that all of the statistical
methods of experimental science can be developed within
our theory, it suffices to show that our Rule of Induction can be used to determine the antecedent probabilities
$q(p)dp$.

This, however, is achieved by a straightforward use
of our rule. We have a sequence of sequences $S: s_1,\ldots,s_i,$
$\ldots,s_k$. We take a certain initial section $s_1,\ldots,s_n$ and
estimate the relative frequency of s_i $(i=1,\ldots,n)$ with
p_i in the interval $p \angle d$. Finally we posit that

92) $q(p)dp = Fn(S,S_{p,d})$

The calculation of $Fn(S,S_{p,d})$, which is required for
this posit, requires that we first posit some values for
$p_1,\ldots,p_n$ (the fact that these are only posited values is

the reason that we spoke of 'estimating' Fn, rather than
calculating it.) These values, too, are obtained by the
use of the Rule of Induction. That is, we examine an
initial section $x_{i_1}, \ldots, x_{i_n}$ of each s_u; calculate $Fn(s_i, B)$;
and posit that

93) $p_i = Fn(s_i, B)$

(Since the sequences are finite, it is also possible
actually to find p_i, $i=1, \ldots, n$, by counting through $s_1, \ldots$
s_n, and thus calculate exactly $Fn(S, S_{p,d})$. This can be
done, however, only when the sequences are not too long.
In this case, only a single posit is required for the
evaluation of $q(p)dp$---namely, 92)).

CHAPTER IV

THE JUSTIFICATION OF INDUCTION

The problem of justifying induction is interpreted by
us, in accordance with the first chapter, as meaning the
problem of showing that success in prediction can be attained by the use of our inductive rule, within the period
in which we are interested in arriving at successful prediction, if it can be attained at all.

The justification we shall give follows, in its broad
outlines, the argument for the infinite case as we presented it in the first chapter. There are, however, several
features peculiar to the finite case. 1) The concept of
'successful prediction' in the finite case requires, as we
shall show, careful analysis, and assumes a somewhat different meaning from the analogous concept for the infinite
case. 2) The concept of 'possibility' of successful
prediction also has to be carefully defined (in the purely formal sense of 'logical possibility' it is evident
that it is always 'possible' to predict correctly every
event in a finite set.) 3) The class of justifiable rules
differs markedly in the finite case from the corresponding class in the infinite case. In order to clarify these
problems, and the general problem of justification, we
shall once more resort to the game-theorteic analogy we

had occasion to use in the first chapter.

In broad philosophical terms, the nature of our task is clear: we are concerned with a finite world of events. We wish to determine the probabilities controlling the worlds. Furthermore, we wish to determine those probabilities in advance, so that this knowledge will have predictive significance.

This final addition, that we wish an early knowledge of probabilities, is that which makes our problem at once significant and difficult. For it is not hard to show that the use of our inductive rule will lead to a determination of the probability controling a sequence, eventually. Indeed, if the sequence is of length n, the final value posited by our rule gives in fact the correct value p=Fn. Not only does our rule give the correct value as we reach the last member of or complete the sequence; it gives values close to the correct value as we come close to the end of the sequence, whatever the nature of the sequence.

Let us suppose, for example, that our sequence is of length n, and that we have examined the first n-m members of our sequence (where we may suppose that m is small in comparison with n). Then the posited value is F_{n-m}. In other words, there are $(n-m) \cdot F_{n-m}$ B's among the first n-m elements of the sequence. Hence, the total number of B's in the sequence lies between $(n-m) \cdot F_{n-m}$ and

$(n-m) \cdot F_{n-m} / m$. In symbols:

1) $\qquad (n-m) \cdot F_{n-m} \angle N \cdot F_n \angle (n-m) \cdot F_{n-m} \neq m$

and therefore

2) $\qquad F_{n-m} \angle \dfrac{n}{n-m} \cdot F_n \angle F_{n-m} \neq \dfrac{m}{n-m}$

If m is sufficiently small in comparison with n, the middle term of this inwquality is approximately equal to F_n, and the right hand is approaximately equal to F_{n-m}, so we have the approximate equality of F_{n-m} and F_n.

For this reason, if all we wish is a rule which yields estimates which approach, and eventually equal, the probability, we need look no further. However, any rule we might use, that is to say, any rule that posits values logically consistent with past experience, has this property, so in this sense induction is not superior to any other procedure we might adopt. Thus, suppose my sequence has ten members, and I have examined the first five and discovered that they are all B's. Any rule which tells me to posit that the probability (in the sense of final relative frequency) is lower than 1/2, is automatically disqualified as giving a result logically inconsistent with the evidence. In general, any rule that tells me to posit, after I have examined the first n-m elements of the sequence, a value outside the interval $\dfrac{n-m}{n} \cdot F_{n-m} \angle p \angle \dfrac{(n-m) \cdot F_{n-m} / m}{n}$ is disqualified as

logically inconsistent. But both the left and right hand
of this interval converge to F_{n-m} as m approaches 0. Thus
the values posited by any logically consistent rule have
the property of converging to Fn which we found for our
rule of induction.

These facts express a fundemental dissimilarity between the finite and the infinite case. In the infinite
case, an initial section of a sequence, however long,
does not preclude the possibility of the limit taking on
any value whatsoever. Hence, no rule can be excluded as
logically inconsistent, in the manner in which we have
been able to exclude many rules in the finite case.

Furthermore, in the infinite case, we do not have
the problem that a rule may give values which finally
approach the limit, but which do not do this until the
end often has been virtually reached. For an infinite
sequence has, of course, no 'end'; and a rule which
gives values which converge to the limit, gives values
which converge early to the limit (if we agree to
regard any finite period, however long, as short in
comparison with the infinite remainder of the sequence).
The problem of early convergence is thus unique to the
finite case. And it is unique to the finite case that the
convergence of the values given by a rule (to the relative frequency) does not, of itself, justify the rule.

While the results we have just mentioned cannot, for this reason, be regarded as constituting a justification of induction, they are, nevertheless, not devoid of significance. For they indicate that the only rules which are admissible (in the sense of logical consistency) are those rules which have the property that the value posited by the rule converges to the relative frequency. Thus the class of justifiable rules of the first chapter corresponds to the class of admissible rules for the finite case[121]. What we wish to do for the finite case is to narrow this class; that is, to show that among these rules there is one, the Rule of Induction, that is justifiable in a stronger sense.

Returning to the terminology of 'games', we may say that our situation is this: we are playing a game in which (as in the first chapter) we deal with sequences of events and our moves are our predictions. But our sequences are now finite, and finite in number. Let us suppose that our sequences are, in fact, of a certain length N'; and that what we wish to do in order to 'win' is to determine by the time we have examined the Nth element of the sequence (for some fixed N depending on the game) what the final relative frequency in the sequence is, within some interval of accuracy d. We may also reasonably demand that our approximation of the final relative frequency should improve as we go on in the sequence[122].

The condition under which our Rule of Induction will
enable us to win is easily formulated: Let us call the
probability p controlling a sequence the 'practical
limit' of the sequence[123] (relative to a fixed N) when the
relative frequency Fn differs from p by less than d in
value, for n=N, and continues to approach p as n approaches
N'. In symbols:

5) $F_N(A,B)-p \leq d$ and $Fn(A,B)-p \leq Fm(A,B)-p$

(where $N \leq m \leq n$).

Then the condition under which we will 'win' with the
Rule of Induction is that the sequence with which we deal
possesses a practical limit. (This follows immediately
from the definition of practical limit, together with the
fact that the values given by the Rule of Induction are, in
fact, the relative frequencies Fn). We shall now show that
the existence of practical limits is the necessary precondition for the possibility of successful prediction.

Before doing this, it is necessary that we consider
an objection that might be raised at this point. So far,
we have identified the problem of prediction with the
problem of finding the probabilities controlling our
sequences. Indeed, we have already shown[124] that these
probabilities, if known, control our 'bets', that is,
our predictions. But it might be asked if 'prediction'
is not also used in another sense; in the sense of finding

a rule which enables us to determine the attribute of the next element in our sequence.

As a first attempt to formulate this sense of 'prediction', we might adopt this definition: a rule R is a good rule for predicting a sequence A if and only if the function f corresponding to the rule R^{125} has the property that

4) $f(n)=0$ if $x_n \epsilon B$
      $f(n)=1$ if $x_n \epsilon \bar{B}$

Let us now define a class C_R as follows: an element x_n belongs to C_R only if it has the attribute B and $f(n)=0$, or it has the attribute $\bar{B}$ and $f(n)=1$. The statement that the rule R is a good rule can now be formulated in this way: every member of the sequence belongs to C_R.

This is equivalent to asserting that $P(A,C_R)=1^{126}$. In this case, we also have that $Fn(A,C_R)=1$ (for all n), and hence the sequence A possesses a practical limit.

Thus, with this definition, the determination that a Rule R is a good rule for prediction is equivalent to the determination of the practical limit of a sequence (a task which can be performed by means of the Rule of Induction).

This is, however, much too restrictive a definition for a 'good rule for prediction'. We come, then, to the problem of formulating more adequately the requirements for such a rule.

One requirement is immediate: if the rule is a good rule, then it must be one which is successful in the long run. That is, the final success-ratio (or $P(A, C_R)$) must be greater than some lower bound k (depending on the game).

Evidently a high success-ratio is not enough, however. For a rule may have a success-ratio close to 1 and still be highly erratic, in the sense of having not only periods of high success, but also periods of continued failure of considerable length. Our dislike of such rules is easily justified. For the predictions obtained by the use of our rule are to guide us in choosing policies; policies on which our welfare and even our existence may depend. A rule which leads to a long sequence of consecutive failure may, for this reason, prove a fatal rule to employ, even though its final success-ratio be high. On the other hand, a rule with a minimum success-ratio, but with great stability, may prove a very good rule to employ. For, while it is true that such a rule only guarantees us a success-ratio of k, we know, when we use such a rule, how many successes and how many failures to expect in any period, and can lay our plans accordingly. The advantage of a stable rule is that it enables us to 'take out insurance' against our failures.

The most natural definition of 'stability' in this sense, is constancy of the success-ratio. Strict constancy is too strong a requirement, however, since $F_1(A, C_R)$

equals 1 or 0 as x_1 is or is not a C_R, and hence the final success-ratio would have to be 1, if we were to demand that $Fn(A,C_R)=p$ for all values of n; we may, therefore, construe the demand as one for relative constancy. That is to say, we ask that $Fn(A,C_R)$ should early assume a value close to $P(A,C_R)$, and should remain approximately equal to it. More simply, what this concept of stability requires is that $P(A,C_R)$ should be the practical limit of the sequence[127].

Thus, to say that there is a good rule for predicting a finite sequence means this: that there is a rule R, which yields a success-ratio greater than k and which is stable. The determination that a given rule is a good rule is thus, once again, a determination of a practical limit of our sequence.

This definition of 'good prediction' leads to two results:

1) The determination that a rule leads to good prediction is a determination of certain probabilities. Thus we are justified in regarding the determination of probabilities as the most general form of the problem of prediction.

2) The existence of a good rule for prediction depends upon the existence of a statistical regularity (that is, a practical limit) in our sequence.

The second point remains valid even if we construe 'prediction' still more broadly, as we did at the outset

of our discussion. Thus, let us take the problem of prediction to be the determination of the probability controlling our sequence. A 'good rule' is now one which,
when applied to a sequence s_i, yields a value R_n which
does not differ from p_i by more than d when $N \underline{/} n$. Again,
such a rule need not be successful for every sequence s_i:
the minimum requirement for 'victory' in our 'game' is that
there be some rule of this sort which succeeds in most
cases (i.e., the success-ratio is greater than some lower
bound k) and which is stable. But, as we have just seen,
this is equivalent to demanding that the sequence S: $s_1, \ldots$
$\ldots, s_i, \ldots$ possess a practical limit. In other words, what
we are now predicting is a sequence of finite sequences,
and what is now required is that this sequence of the
second level possess a practical limit.

The same discussion can be carried out on any level
of prediction. The point that we establish in each case
is the same: namely, that to say that there is a good
rule for predicting a sequence is already to assert the
existence of a certain kind of statistical regularity in
that sequence.

We are now able to apply the justification of induction offered in the first chapter, to the finite case. In
the finite as in the infinite case, this is its form: what
we are interested in finding are statistical regularities.
Such regularities (limits) must exist, indeed, if predict-

ion is possible at all. We cannot show that they exist; but we employ the Rule of Induction to find them because we know that it will lead to a correct determination of these limits _if they exist._

Once again, the proof of the assertion that the Rule of Induction will determine the value of the practical limit of a sequence, if it exists, follows immediately from the definition of the limit-concept. For, by a 'determination of the value' we understand the giving of a series of estimates $R_1, R_2, \ldots$ such that R_n differs from the true probability by less than a fixed amount d, whenever $N \underline{/} n$. In the case of the Rule of Induction, this series is simply $F_1, F_2 \ldots$; and the statement that these estimates constitute a determination of the value in this sense, that is that F_n differs from F_N by not more than d whenever $N \underline{/} n$, is also the statement that the sequence possesses a practical limit. This, then, is the essence of the justification: the statement that a sequence possesses a practical limit is logically equivalent to the statement that the continued use of the Rule of Induction will successfully determine the value of that limit.

While this justification is identical with that given by Reichenbach for the infinite case (and the possibility of adapting that justification to the finite case in this way has been pointed out by Reichenbach in his Theory of Probability[128]), the steps leading up to

that justification differ in one important respect from
the treatment given by Reichenbach; namely, we do not i-
dentify 'probabilities' in the finite case with 'practical limits', but, more broadly, with 'final relative
frequencies'. It is for this reason that we become involved in a problem that has no analogue in the infinite
case; the problem of showing that the determination of
probabilities can be reduced to the determination of practical limits. What we have had to show, accordingly, is
that the determination of the probabilities controlling
our sequences, or, more generally, the prediction of our
sequences, demands the existence of practical limits. The
probability P(A,B) can often be determined even when it is
not a practical limit; but it is necessary that some other
probability from which it can be computed be a practical
limit, either of the sequence itself or of a sequence on a
higher level.

Our reason for not identifying probabilities with
practical limits are fundamentally mathematical in character. For the Rule of Existence demands that, if the probabilities which uniquely determine it exist, a probability
exist. Thus, if P(A,B) exists, and P(A,C) exists, and
B and C are incompatible, then P(A,BVC)=P(A,B) $\neq$ P(A,C)
must exist. But a sequence may possess a practical limit
with regard to the relative frequencies Fn(A,B) and
Fn(A,C), but not possess a practical limit with regard to

Fn$(k, \text{BVC})^{129}$. As a consequence we are forced to reject
(as inadmissible) 'practical limit' as an interpretation
of 'probability'.

In point of fact, our interpretation of 'probability'
is completely forced upon us, as we have shown, by the
axioms we have accepted. For it is a consequence of these
axioms that 1) every finite sequence is a probability
sequence, and 2) the probability controlling it is simply
the final relative frequency.

On philosophic grounds, also, we are inclined to prefer the more lenient definition of 'probability'. For the
final relative frequency in a sequence controls our bets,
as we have pointed out, whether or not it is also a practical limit. Furthermore, probabilities in finite sequence
can often be determined even when they are not practical
limits (because they depend mathematically upon other
probabilities which are practical limits.)

We must also consider the problem of what it means to
say, in the finite case, that good prediction is 'possible'.
As we have stated above[130], it is always possible to predict a finite sequence if by 'prediction' we understand
merely the discovery of a recursive rule defining the sequence. (For any finite sequence may be defined by a
recursive rule.) However, it is not simply a recursive
rule, or a calculable function[131], that we seek, but a
practically calculable function. The fact that a function

is in principle 'effectively calculable' does us little
good if we are unable to actually carry out the calculation, or to build a machine that will do it; and, similarly, it does not help us to have a 'recursive rule' if
we cannot determine the prediction that follows from the
rule in any specific case. In view of the great length
of our sequences, we cannot give any proof on apriori
grounds that there exists a practically recursive rule, in
this sense, for predicting them. But what we have shown is
that, if such a rule exists, then there exist precisely
defineable attributes C_R such that the probabilities
$P(A,C_R)$ are practical limits of these sequences.

We don not believe, in other words, that it is <u>a</u>
<u>priori</u> that successful prediction is possible, in any
significant sense. But we hold, as we did in the infinite
case, that it is not necessary to prove this in order to
justify induction.

Finally, we must ask whether there are not other
rules that are justifiable in the same sense as the Rule
of Induction. In the infinite case, we saw that every
rule which gives values which converge to the relative
frequency was also justifiable. Similarly, in this case,
every rule which yields values which, after the point N,
differ from the practical limit by not more than d, and
which finally converge to the practical limit, is justifiable in the same sense as the Rule of Induction; it may

be said of any such rule that if practical limits exist,
it will enable us to find them.

But the class of such rules is severely limited. Indeed, any such rule must give values Rn which are equal to
the relative frequencies Fn wheneber $N \angle\angle n$. For, let us
suppose that a rule gives a value R_n different from Fn. It
is possible (since we suppose that the values given by the
rule depend solely on the initial section $x_1, x_2, \ldots, x_n$;
and hence the remainder of the sequence is free to behave
in any way, subject only to the restriction, when the sequence has a practical limit, that F_j must remain in the
interval Fn$\angle$d---since both F_j and Fn must be in the interval p$\angle$d) that the final relative frequency is Fn-d (if
Fn is smaller than Rn), or Fn$\neq$d (if Rn is smaller than Fn).
But in this case the difference Rn-p would be greater than
d.

That is, if R_n differs from F_n ($N \angle\angle n$), it is possible to find a sequence possessing a practical limit on
which the rule R fails.

We may sum up the results which we have obtained in
the following manner: let us represent the class of possible rules by

Fn $\neq c_n$

Then a rule is <u>admissible</u> if $c_n \rightarrow 0$ as $n \rightarrow N'$; but it
is <u>justifiable</u> only if $c_n = 0$ when $N \angle\angle n$. If we agree to

ignore the first N values given by a rule, we may say
simply that the only justifiable rule is the Rule of
Induction.

As a particular consequence of this, we have that
Carnap's probability function c^* is not a justifiable
rule[132]. For c^* is, in general, not exactly equal to the
relative frequency unless the whole sequence has been
examined, though it converges to the relative frequency.
Hence, we cannot say that c^* will equal the probability
controlling the sequence (within the degree of approximation d) whenever we have passed the point N in our examination of the sequence, even though the sequence possesses a practical limit[133].

NOTES AND REFERENCES

Chapter I

1. For a criticism of this formulation, see: Arthur Pap,
Elements of Analytic Philosophy (New York: Macmillan,
1949), pp. 402-7.

2. The Pre-Socratics, in particular, stressed the non-
veridical nature of sense-perception. According to
them, the fact that the senses give different reports
to different observers, and the existence of illusion,
indicates that one can never know that the report of
one's senses is correct. Modern scepticism, on the
other hand, while agreeing that we can never infer
from sense-data to the material world, has held at
least one thing to be immediately certain---the felt
quality of the experience itself. This was never
regarded as significant by the ancients; hence, they
never attempted to construct knowledge on a phenome-
nalistic basis.

3. Francis Bacon, Novum Organum (New York: Willey, 1944).

4. Pap, op. cit., 362-8.

5. For reports of immediate experience, while they constitute the epistemological foundation for science,
form a very small part of the corpus of scientific
knowledge. Furthermore, even 'report-propositions',
like 'I see a chair', go beyond immediate experience.
Cf. Hans Reichenbach, Experience and Prediction
(Chicago: University of Chicago, 1949), pp. 85-7.

6. That is, the Pre-Socratics, again. Their arguments
resemble Hume's argument against any justification
of induction: to prove any logical principle (they
argued) we must assume either the same principle
or others---which must then be proved. Hence, the
process of proof is either circular, or leads to an
infinite regress.

7. Bertrand Russell, A History of Western Philosophy
(2nd ed.; New York: Simon and Schuster, 1945), p.832.

8. An Enquiry Concerning Human Understanding (La Salle:
Open Court, 1907), Book I, Part IV, Sec. I.

9. Introduction to Aristotle (New York: Modern Library,
1947), p. 15.

10. Op. cit., 673-4.

11. Ibid., 668.

12. While these game-theoretic concepts were developed by Von Neumann within probability theory, we apply them here to the foundations of the theory, as purely logical concepts.

13. The Theory of Probability (2nd ed.; Berkeley and Los Angeles: University of California, 1949), pp. 469-82. Cf. Experience and Prediction, Chapter V.

14. The Theory of Probability, 347-9.

15. Collected Papers (Cambridge: Harvard, 1936), V, pp. 365-70.

16. Reichenbach, loc. cit.

17. A convenient term, suggested to me by Reichenbach.

18. Russell, op. cit., 670.

19. Laplace, Essai Philosophique sur les Probabilites (Paris: Gauthier-Villars, 1921), p.9.

20. Peirce, Venn, Reichenbach, Carnap, etc.

21. Throughout this work the much abused term 'empiricism' denotes simply the rejection of synthetic a priori principles. In other words, an empiricist is one who believes that statements of fact can only be confirmed by reference to experience.

22. J.M. Keynes, A Treatise on Probability (London: Oxford, 1921); Carnap, The Logical Foundations of Probability (Chicago: University of Chicago, 1950); O. Helmer and P. Oppenheim, "A Syntactical Definition of Probability and of Degree of Confirmation," Journal of Symbolic Logic, X (1945), p. 25.

23. C. Stumpff in an article published in Ber. d. bayer. Akad. philos. Kl. (1892). For a discussion of this view, see: Reichenbach, The Theory of Probability, pp. 368-9.

24. D. Hilbert and P. Bernays, Grundlagen der Mathematik, Vol. I (Berlin: Springer, 1934), pp. 1-30.

25. The Theory of Probability, chapter III.

26. This interpretation is 'purely mathematical' in that no empirical content is assigned to probability by it. In this respect it is comparable to the analytic interpretation of geometry.

27. A. N. Kolmogoroff, Grundbegriffe der Wahrscheinlichkeitsrechnung (Berlin; 1933).

28. Probability, Statistics, and Truth (New York: Macmillan, 1939).

29. The Theory of Probability, 72-5 and 138-9.

30. Ibid., 69.

31. Ibid., 45.

32. Certain interpretations of probability apparently differ from us in regarding probability as a property of event-propositions, rather than the events themselves. Our theory may be easily reinterpreted in this way, however (see: The Theory of Probability, chapter X.)

33. Pogo Possum, a comic book (Los Angeles: Dell, 1951).

34. Introduction to Aristotle, 293.

35. We think of a rule R as yielding a series of esti-mates $R_1, R_2, ..., R_n, ...$ Thus, the variable n denotes the number of the estimate.

36. That is to say: pick any interval of approximation d. Then, when the values given by B differ from the values given by A by not more than $\frac{1}{2}d$, and when these in turn differ from the true values by not more than $\frac{1}{2}d$, the values given by B will differ from the true values by not more than d.

37. Reichenbach, The Theory of Probability, 338.

38. Ibid., 430.

39. Since a probability implication may refer to any number of sequences, we should say 'or with regard to an n-tuplet of infinite sequences'.

40. Ibid., 444-51.

41. Ibid., 429.

42. Ibid., 373.

43. Ibid., 379.

44. Ibid., 479.

45. Ibid., 481.

46. See page 6.

47. Because, to say that a series of values will event-ually differ from a fixed number by less than d, no matter what positive number we may choose for d, is (by definition) to say that the series approaches the fixed number as its limit.

Chapter II

48. That the use of probability as a guide to action must be justifiable.

49. Grundlagen der Mathematik.

50. Topology (New York: Harpers, 1938).

51. Page 15.

52. Cf. The Theory of Probability, 49-52.

53. For the sake of simplicity, we assume a type-free system of logic. For our purposes, all that will be required is some arithmetic and quantification theory.

54. The degree depends upon whether we wish to speak of a sequence of events (degree one); a sequence of sequences of degree one, or two-dimensional lattice (degree two); a sequence of such lattices....; etc.

55. Op. cit., 173.

56. Pages 19-20.

57. Reichenbach, op. cit., 409.

58. That is, if single-case statements like 15) were ad-mitted to the system, and the Principle of Univocali-ty were extended to them, inconsistency would result.

121

59. The Theory of Probability, 49-52.

60. Meaning and Necessity (Chicago: University of Chicago, 1947), p. 46.

61. The principle, as formalized, is actually stronger than this rendition of it.

62. Meaning and Necessity, 53. We use 'intensional' to mean simply 'non-extensional', not in Carnap's sense.

63. Op. cit., 53.

64. For example, The Theory of Probability, 51.

65. Where the substitution is of corresponding formulae for corresponding variables; each formula being put for all the occurences of the corresponding variable (i.e., the substitution is uniform).

66. The c_i need not be distinct.

67. The Theory of Probability, 132-5.

68. Cf. ibid., 135.

69. Probability, Statistics, and Truth.

70. Alonzo Church, "Randomness", Bull. Amer. Math. Soc., XLVI, (1940) p. 130.

71. Loc. cit.

72. Loc. cit.

73. Following Reichenbach, op. cit., 148.

74. Ibid., 138.

75. Cf. ibid., 167.

76. In his course in Mathematical Logic (Math. 258) at Harvard University.

77. His Mathematical Logic (2nd ed.; Cambridge: Harvard, 1948), represents a system developed along these lines.

78. Op. cit., 143-51.

79. "Admissible Numbers in Probability", Amer. Jour. Math., 50, No. 4 (1948), p.535.

80. The Theory of Probability, 63.

81. Ibid., 136-7.

82. Page 60.

83. Page 48.

84. Cf. Reichenbach, op. cit., 142.

85. Ibid., 63.

86. I.e., assuming that A is not empty. If it is empty, the Special Theorem of Multiplication may of course be used (since the probability takes on any value).

87. Cf. Reichenbach, op. cit., 144.

88. Ibid., 145.

89. Loc cit.

90. Page 58.

91. See page 13.

92. By the same method, we may show that, in an infinite periodic sequence, the probability is equal to the limit of the relative frequency (or the relative frequency in any period). The proof is obtained by replacing the c_i by the regular divisions s_{ni} (where n is the period of the sequence) in the proof given.

93. Reichenbach has proposed this term to replace the more usual 'apriori probability'---which has unfortunate metaphysical connotations. Op. cit., 93.

Chapter III

94. The term 'primitive knowledge', as used by Reichenbach, denotes the situation in which we have no inductive knowledge (in particular, no knowledge of probabilities of a higher level); 'advanced knowledge' is the term for the situation in which we do possess such information. Op cit., 364.

95. With a very small degree of approximation d.

96. In practice, we deal with small samples and large N.

97. This is the 'lattice' interpretation of this question.

98. E.g., by using Stirling's Formula to approximate the factorials. See, in this connection, Probability, 270.

99. Loc. cit.

100. Loc. cit.

101. Cf. Reichenbach, Probability, 274.

102. By the mean of a sequence to which is coordinated a finite set of amounts u_k we understand the sum of the products obtained by multiplying each amount by the probability of getting that amount in the sequence. This is the 'theoretical' definition of the mean.

103. Since the nth member of the sample, taken for each sample, runs over all the members of the sequence (because of our circular counting), for every n.

104. See Reichenbach, op. cit., 194.

105. Ibid., 191-2.

106. Ibid., 275.

107. Loc. cit.

108. Loc. cit.

109. And constitute a normal lattice. For the definition of this, see Reichenbach, Probability, 172.

110. Cf. Probability, 326.

111. Because a probability with a small interval in the reference class is approximately equal to the probability calculated for an exact value (in that interval) in the reference class.

112. Cf. Probability, 93.

113. Ibid., 172.

114. Ibid., 306.

115. The Bernoulli Theorem.

116. See Probability, 328

117. Ibid., 306.

118. We suppose N to be very large, so that this is not a very serious restriction.

119. probability, 335.

120. The use of the Rule of Induction to determine the normalcy of sequences is explained by Reichenbach, op. cit., 463-4.

Chapter IV

121. More precisely: In the case that all of our sequences are of length N', a rule is admissible only if it converges to F_N. When we do not know the length of our sequences, the only rules that can be excluded are those which do not ever converge to the relative frequency.

122. This demand does not play any significant role in the development of the theory, however.

123. See Probability, 347.

124. Page 58.

125. A function f corresponds to a rule R if f(n) equals 0 or 1 as R predicts x_n to be a B or a $\bar{B}$, respectively.

126. In the finite case, probability 1 has the meaning of 'no exceptions'. This is an important difference from the infinite case, where probability 1 is compatible with an infinite number of exceptions.

127. We might weaken the requirement of stability still more, and merely demand that the number of failures in any period of r successive predictions should not exceed s. In this case the sequence itself may not possess a practical limit---but the coordinated frequency sequence (the sequence of relative frequencies of C_R in the set of r successive members of the sequence beginning with x_i---i=1,2,...,n) possesses a practical limit.

128. Op. cit., 347.

129. For the interval of convergence may be doubled.

130. See the beginning of this chapter.

131. A. M. Turing, "Computability", Jour. Symb. Logic, 11, no. 1 (1946), p. 153.

132. It is, of course, justifiable in a weaker sense, for
while it does not determine the final frequency
when we have reached the point N (within the degree d),
it almost determines it (that is, it determines it
within a slightly larger degree of approximation). We
may also say that it determines it slightly later than
N. In general, a rule is justifiable to the degree
that it corresponds with the relative frequencies I_n
after the point n=N.

133. This is of great practical significance, for in
the actual application of induction we always hope
that the sequences we treat possess a practical
limit. Hence, we prefer the rule that converges most
rapidly to it, when it does exist.

BIBLIOGRAPHY

Aristotle. *Introduction to Aristotle*. Edited by Richard Mckeon. New York: Modern Library, 1947.

Bacon, Francis. *Advancement of Learning and Novum Organum*. New York: The Willey Co., 1944.

Carnap, Rudolph. *Meaning and Necessity*. Chicago: University of Chicago, 1947.

Church, Alonzo. "Randomness," *Bulletin of the American Mathematical Society*, XLVI (1940), 130.

Copeland, A. "Admissible Numbers in the Theory of Probability," *American Journal of Mathematics*, 50 (1948), 535.

Hilbert, David, and Paul Bernays. *Grundlagen der Mathematik*. Berlin: Springer, 1935.

Hume, David. *An Enquiry Concerning Human Understanding*. La Salle: Open Court Publishing Co., 1907.

Keynes, John Maynard. *A Treatise on Probability*. London: Oxford, 1921.

Kolmogoroff, A.N. *Grundbegriffe der Wahrscheinlichkeitsrechnung*. Berlin: Springer, 1933.

Laplace, H. *Essai Philosophique sur les Probabilites*. Paris: Gauthier-Villars, 1921.

Pap, Arthur. *Elements of Analytic Philosophy*. New York: MacMillan, 1949.

Peirce, Charles Saunders. *The Collected Papers of Charles Saunders Peirce*. Cambridge: Harvard University, 1934.

Reichenbach, Hans. Elements of Symbolic Logic. New York:
    Mac Millan, 1947.

----------Experience and Prediction. 3rd ed. Chicago:
    University of Chicago, 1949.

----------The Philosophic Foundations of Quantum Mechanics.
    Berkeley and Los Angeles: University of California,
    1948.

----------The Theory of Probability. 2nd ed. Berkeley and
    Los Angeles: University of California, 1949.

Russell, Bertrand. A History of Western Philosophy. New York:
    Simon and Schuster, 1945.

----------Human Knowledge, its Scope and Limits. New York:
    Simon and Schuster, 1948.

----------An Inquiry into Meaning and Truth. New York:
    W.W. Norton and Co., 1940.

Todhunter, I. A History of the Mathematical Theory of Proba-
    bility from the Time of Pascal to that of Laplace.
    New York: G.E. Stechert and Co., 1931.

Von Mises, Richard. Notes on the Mathematical Theory of Pro-
    bability and Statistics. Special publication no.
    1. from the Graduate School of Engineering,
    Harvard University, Cambridge, Mass. (mimeographed).

----------Probability, Statistics, and Truth. Translated by
    J. Neyman, D. Scholl, and E. Rabinowitsch. New
    York: MacMillan, 1939.

Williams, Donald. The Ground of Induction. Cambridge, Mass.:
Harvard University, 1947.

For Product Safety Concerns and Information please contact our EU representative GPSR@taylorandfrancis.com Taylor & Francis Verlag GmbH, Kaufingerstraße 24, 80331 München, Germany

Printed and bound by CPI Group (UK) Ltd, Croydon, CR0 4YY

01/05/2025

01858520-0002